The Second Renaissance

Reclaiming Prosperity in a Decentralized Age

PUBLISHED BY Mike Hobart

The Second Renaissance

Copyright © 2025 Mike Hobart

All rights reserved.

ISBN-13: 979-8-9929470-0-7

No part of this publication may be copied, reproduced in any format, by any means, electronic or otherwise, without prior consent from the copyright owner and publisher of this book.

Table of Contents

The Second Renaissance 1
Acknowledgements 9
    Introduction 13
    Part 1

Setting the Board 15
    CHAPTER 1: The Problems 16
    CHAPTER 2: Education 18
        How Did We Get Here? 19
        Moats & Ivory Towers 20
        Eroding Our Future 22
        Who Pays The Greatest Price? 26
    CHAPTER 3: The Social Contract 29
        Civilizations And Health 30
        Earth and Empire 31
        Addictions 33
        The Problem 35
    CHAPTER 4: Economics & The World Around Us 40
        Subtle Diffusion 40
        On Feminism 41
        Psychology's Power Over Physiology 42
        A Developmental Question 44
        Home & Parental Environment 46
        An Honorless Money 50
        The Divine Masculine 51
        The Divine Imperative 52
        A Detour Into Cringe 53
        Mountains of Debt & Doubt 55

The Second Renaissance

Part 2

.... 57

Value Exchange 57

CHAPTER 5: An Uncommon Look at Common Economies . 58

CHAPTER 6: A New Way of Looking 60

    Let There Be Light 62

    Checks & Balances 62

    The Economics of Life 63

    Competition & Progress 64

CHAPTER 7: Economies of Complex Systems 66

    The Economics of a Body 67

    Carbs, Fats, Calories & ATP 68

    Inflammation in Biological Economies 68

    Inflammation in Human Economies 70

CHAPTER 8: A Tangled Mess We Weave 71

CHAPTER 9: Let's Talk Energy 73

    Oil & Energy 74

    Fossil Fuels & Powering the World Economy of Value 74

Part 3

.... 76

The Second Renaissance 76

CHAPTER 10: Where Did the Time Go? 77

CHAPTER 11: Breaking The Paradigm 79

CHAPTER 12: A Spoonful of Liberty… 80

CHAPTER 13: Stress Relief 82

CHAPTER 14: Free Time. Free Spirit. 84

CHAPTER 15: Valued In Absence 85

CHAPTER16: Revolutionary Momentum 87

CHAPTER 17: History's Lesson in Today's Flavor 89
CHAPTER 18: The Incipient Energy Revolution 91
    Emergent Catalysts 91
    Economic Constraints and Bitcoin's Role 93
    Societal Transformation 95
CHAPTER 19: Reviving The American Dream 96

Part   4

.... 98

    The Energy Renaissance 98
CHAPTER 20: Politicization Of Energy 99
CHAPTER 21: Energy and Growth 103
    Energy And Return On Investment 106
    The Solow Model 107
    The Solow Model & Energy 108
CHAPTER 22: The Future Of Energy: Bitcoin Mining 110
    The Future Of Energy 111
    Conclusion 114
CHAPTER 23: National Defense Through Bitcoin Mining ... 115
CHAPTER 24: Decentralized Defense 119
CHAPTER 25: Ad Valorem 121
CHAPTER 26: Separating Money from the State 123
CHAPTER 27: The Looking Glass & The Sword 126
Chapter 28: Bitcoin Enables Fiat Survival 128
    A Blackhole for Excess Liquidity 129
Chapter 29: Peaceful, Not Harmless 132
    OPN Bitcoin 132
    The Brave Mission 133
    Helene and North Carolina 135

Our BVs..137
The Orange Sun Rises..........................139
Thank You.....................................141
REFERENCES:...................................142
The Second Renaissance..........................1
  Introduction.................................10
  Part..1

Setting the Board..............................12
  CHAPTER 1: The Problems......................13
  CHAPTER 2: Education.........................15
    How Did We Get Here?.......................16
    Eroding Our Future.........................19
    Who Pays The Greatest Price?...............23
  CHAPTER 3: The Social Contract...............26
    Civilizations And Health...................27
    Earth and Empire...........................28
    Addictions.................................30
    The Problem................................32
  CHAPTER 4: Economics & The World Around Us...37
    Subtle Diffusion...........................37
    Psychology's Power Over Physiology.........39
    A Developmental Question...................41
    Home & Parental Environment................43
    An Honorless Money.........................47
    The Divine Masculine.......................48
    The Divine Imperative......................49
    A Detour Into Cringe.......................50
    Mountains of Debt & Doubt..................52

Mike Hobart

Part 2

.... 54

Value Exchange 54

CHAPTER 5: An Uncommon Look at Common Economies 55

CHAPTER 6: A New Way of Looking 57

    Let There Be Light 59

    Checks & Balances 59

    The Economics of Life 60

    Competition & Progress 61

    CHAPTER 7: Economies of Complex Systems 63

    The Economics of a Body 64

    Carbs, Fats, Calories & ATP 65

    Inflammation in Biological Economies 65

    Inflammation in Human Economies 67

CHAPTER 8: A Tangled Mess We Weave 68

CHAPTER 9: Let's Talk Energy 70

    Fossil Fuels .... **Error! Bookmark not defined.**

    Oil & Energy 71

    Fossil Fuels & Powering the World Economy of Value 71

Part 3

.... 73

The Second Renaissance 73

CHAPTER 10: Where Did the Time Go? 74

CHAPTER 11: Breaking The Paradigm 76

CHAPTER 12: A Spoonful of Liberty................................ 77

CHAPTER 13: Stress Relief 79

CHAPTER 14: Free Time. Free Spirit. 81

CHAPTER 15: Valued In Absence 82

CHAPTER 16: Revolutionary Momentum 84
CHAPTER 17: History's Lesson in Today's Flavor 86
CHAPTER 18: The Incipient Energy Revolution 88
    Emergent Catalysts 88
    Economic Constraints and Bitcoin's Role 90
    Societal Transformation 92
CHAPTER 19: Reviving The American Dream 93
Part ———————————————————————— 4
.... 95
The Energy Renaissance 95
CHAPTER 20: Politicization Of Energy 96
    Energy And Return On Investment 103
    The Solow Model 104
    The Solow Model & Energy 105
CHAPTER 22: The Future Of Energy: Bitcoin Mining 107
    The Future Of Energy 108
    Conclusion 111
CHAPTER 23: National Defense Through Bitcoin Mining 112
CHAPTER 24: Decentralized Defense 116
CHAPTER 26: Separating Money from the State 120
CHAPTER 27: The Looking Glass & The Sword 123
Thank You 138
REFERENCES: 139

Mike Hobart

Acknowledgements

I want to thank Melanie and Gio for so readily accepting me into your lives, I know I'm not the easiest individual to live with, and I look forward to building our family together. You two inspire me to be better than I was yesterday.

I cannot thank my parents and sister enough. Mom, you taught me (painstakingly) the skills necessary for more accurate and precise communication, analysis of people and their behaviors, and how to control my temper and frustration with the rest of the world that fails to keep up with me. Anna, I know I haven't been the best brother to you (I still feel bad for tricking you out of your Charizard as kids), but I hope to be a more present and proper brother over the years as we both grow our own families. Dad, you taught me some of the most valuable and useful skills that have allowed me to flourish in my time as a veteran—the importance and impact of a true team player, how to foster loyalty amongst your peers by leading by example, and how to do the job right first so that the job can be done with speed and precision later. Those behaviors and skills are core to the reputation I have built today, one which I am proud of. All of that is thanks to having a successful family unit. And I will never claim that to be something as ridiculous as 'white privilege,' because it was thanks to the hard work of my parents.

Over these years I have been blessed with friendships with easily the most interesting people I have ever had the pleasure of crossing paths with. I met all of them in some form of online interaction and culminated in such deep sharing of ideas, theory, and general

The Second Renaissance

conversation that we ended up meeting in-person and our friendships flourished by orders of magnitude since. The closest of which are friendships that I hold dearest, those who I view as mentors that have-, and continue to-, contribute to my development as a man, a warrior, a sheepdog and a writer. So I want to say thank you. Thank you to: Shane Hazel, Jordan Gambrell, Gabe Lord, Alex Stanzyck, CJ Wilson, Lisa Hough, Colin Crossman, and TexasSlim.

I want to extend a special thank you to Amanda Cavaleri for being a powerful inspiration and for introducing me to individuals that changed my life forever. Wherever you are, and whatever you are doing, I hope your explorations–in all things–provide you what it is you seek. I trust–and hope–that you remain safe in your travels.

I want to thank the members of my former Great American Mining (GAM) team that I had the distinct pleasure of working with, albeit a brief amount of time. Thank you, Todd Garland, you took a risk on me, and were one of the best bosses I've had. I hope it paid out for you as you had hoped! And I appreciate our periodic check-ins, I hope we can meet for dinners and coffees in our future. Thank you to the muscle behind GAM: Brad Cuddy, Tyler Hunt, Michael VanDeventer, and Tyler Schmill. I know that the Oil & Gas space is a club of men with shared experiences, one that I am still not a part of, but was welcomed into the organization as a member of the team. As a soldier, I understand how that may feel, helping me appreciate it all the more.

My appreciation for the inclusion into the O&G club of friendships extends to Dan Morrison, Steve Barbour, Adam O., Jonathan Kohn, Jake Corley, and Justin Ballard. I had a great time during my short stint in Houston, TX, where I learned much and am very happy to have made those memories.

I want to thank the Simply Bitcoin team: Opti, Nico, and Rusty. Thank you for giving me an outlet to rant and rave with a few of the absolute best, and for simply being good friends. I hope the best for all of you, and I look forward to continuing to work alongside you and watch you all grow over the years.

Thank you to the Bitcoin Magazine team for being among the first to put weight behind my work as a writer, for being a part of my refinement as a creator, and for bringing me together with so many inspiring and daunting individuals. These men and women include: CK, Shinobi, Mark Goodwin, Joakim Book, L0laL33tz, Tyler Leroche, Dylan Leclair, and Joe Rodgers. You're all among the best and readily willing to challenge ideas and engage in the intellectual sparring that fosters growth and improvements in understanding that is so rare in the world.

I want to thank Jay Gould, for being one of my earliest friendships that I made on Clubhouse. You were the first interaction that catalyzed my fascination in all things macroeconomics and geopolitics with our 5AM discussions on bond markets while the rest of the world avoided starting their day. I'm sorry for your nightmare of a medical situation you went through, and I can say I am very happy you are still with us today. I know your family feels the same way.

Continuing with my fascination in the macro, which has changed the ways that I view the world for the better, I want to thank: Larry Leppard, Dr. Jeff Ross, Preston Pysh, Magoo, Jason Lowery, and Robert Breedlove.

Thank you to my friends Matthew Parker and Emma Muhleman. You are two friends that are specifically outside the Bitcoin space that I hold in high regard, and value deeply. Our conversations rank among my favorites and I will always look forward to catching up.

Thank you to Heidi Dohlman. You were the first individual that I have great respect for that pushed me into starting writing–I'm happy you did, and that I followed through. You rank among my favorite people, and I thank you for your friendship and your words over the years. Oftentimes they were precisely what I needed to hear.

Thank you to all of our Bitcoin Veterans. You boys have built a comradery that I would never have envisioned in my future. We are building something truly great just through our friendships alone. We are a network that has power and will grow to become more powerful than anyone–including ourselves–could have ever imagined.

The Second Renaissance

This book is dedicated to my son. You will be joining us soon, and I hope I can provide you with the example and skills that will develop you into the man you deserve to be—

a King among~~st~~ his peers.

Introduction

It's time that a lot of folks come to that difficult realization—that long standing institutions have festered and failed all of us. That legacy media, academia, science, health, and military leadership have all failed all of us. We have hit a moment in history where the very pillars of our society must be reimagined and retooled to continue supporting that which is built atop them. Otherwise what is no longer of use must be torn down to make room for what is best-fit.

Welcome to the age of decentralization. The pendulum of progress is gaining momentum in a direction opposite of the trajectory it has held for over a century. Consolidation and centralization brought many advantages, but it is not a panacea. We have run that course to its end. Now we begin the next phase of the journey.

In this phase much confusion and chaos will occur. This is the portion of time in the pendulum's swing where it seems to float on air as the momentum redirects. As time goes by, the direction and path become clearer. Resources and forces reorganize into a path of least resistance (which is *often*—not always—the path of greatest efficacy). With it, the fear and confusion will subside. Predictability will return. A "normal" will get established once again.

Decentralization will return rights to the individual, rather than in the hands of government bureaucrats who are not experts in anything beyond bureaucratic congestion and inefficient action. Doctors and teachers will be allowed to be REAL doctors and educators, not the big pharma sales reps and propagandists that they have become. Innovation will reignite, especially as energy generation becomes more effective and diversified. Economics will allow the businesses with the best products, services, and economics to gain their rightful place within The Market.

None of it will be perfect, in execution or in efficacy. But it will bring improvements from where we are now. Progress is progress. We cannot hinder advancement because it is not perfect, when nothing in existence can be.

The Second Renaissance

I do not intend for the content of this publication, or the work cited, to be interpreted as works of my own. The only service I am providing is to hopefully arrange these details in a manner that is easily digestible so that you can be as awestruck as I was once the dots connect. After which I will share what I believe to be the solutions that will yield enriching environments. Environments that will best enable further solutions. Positive feedback loops. Efficiency. Solutions that provide solutions.

This publication is intended to tie the bow off on the question to "Why Bitcoin?" We will be exploring the possibilities for the (bitcoin mining + energy) relationship and its justification for widespread adoption and integration into power production & distribution grids the planet-over while also relating to what is enabled when it couples with bringing back saving as a wealth generating strategy thanks to bitcoin.

First, I want to identify some of the problem areas.

Mike Hobart

Part 1

Setting the Board

CHAPTER 1: The Problems

Welcome fellow time traveler to… The Future. We do not have flying cars. We do not have energy weapons. We have not visited The Moon since the 1960s. The world has been embroiled in war since the 'war to end all wars' had ended in victory for The Allied Forces in the 1940s.

At the time of publishing the populations of the richest and most modern developed nations are plagued by a catalog of chronic health conditions. These same populations are heavily addicted to the products of massive pharmaceutical conglomerates that numb life's experiences. Many others are relying upon amphetamines to compete in the employment and academic markets thanks to socialist policies like Diversity, Equity and Inclusion (DEI) initiatives – which throw all consideration of merit for employment and rely solely on the color of one's skin, their nationality, their religion, or their sexual orientation.

With the advancements in technology enabling the free exchange of information in very rapid and diverse manners, we were hopeful that the vast majority of the public would then have the power to go off and produce geniuses everywhere. What has resulted has been anything but. We have swaths of males and females addicted to pornography and video games. Which we have learned kick off potent dopamine stimulation in a combination: of blue light from the devices' screens and the sheer exposure to the content that is delivered, whether it is sexually stimulating, or exciting, or emotionally gripping—a potent combination. This amalgamation hijacks the wiring of our bodies that enabled our species' successes in a brutal and merciless natural world.

Along with these advancements in technology we have built a world where laborious work is in less demand from the average citizen, *but it is not in zero demand*. Labor has become a scarce resource, as the average citizen is arguably in the worst physical shape that the average male or female has ever been in history. In 2024 diabetes alone affected 12% of the adult US population as being *diagnosed*, with 23% of diabetics being **undiagnosed**, and an additional 38% of the adults in the US being **prediabetic**[1].

Technological advances have provided us with a source of energy that yields potent and very reliable power; nuclear energy. Yet, groups within the population spread unreasonable fear over the technology due to lessons learned along the way (Chernobyl and Fukushima). Now we focus our efforts primarily on energy sources that are not reliable, are not potent, and we even have meaningful percentages of the population claiming the solution is <u>not</u> to produce more energy, but to reduce our energy *consumption*. When human flourishing is directly tied to the production and utilization of greater and greater amounts of energy, as argued by Alex Epstein in his publication *Fossil Future*.

With the overproduction of unreliable and impotent forms of energy generation, we get expensive power production. With expensive power production we get expensive production of goods and expensive provisioning of services. These problems end up swinging right back around to power production like a smart missile that was distracted from its target swooping back around in a wide arc to get back on-target. The power infrastructure of the United States is in a state of decay, our own American Society of Civil Engineers (ASCE) dishes out a report card on the status of our great nation with grades on each individual industry and aspect of infrastructure. In 2021 the ASCE gave our country's grade an overall (C-), as well as our energy sector[2]. A C-MINUS.

---

[1] https://www.cdc.gov/diabetes/data/statistics-report/index.html
[2] https://infrastructurereportcard.org/

CHAPTER 2: Education

How do we define "education"? Is it something that can only be provided by a structured institution or certified individual (such as a teacher, or professor)? Do our parents and friends not teach us? Is it something an individual builds? Or is it something one receives? Do we discover education? Do we mold it? Is it something that we grow? It isn't tangible, yet, can you not receive it, exchange it and develop it?

Education does not exist within the world of atoms, nor within the plane of bits. Yet, it is so real that we can construct an education, we can gift an education to others, and it can be fostered to grow and develop.

Education has been one of the most important pillars of civilization and has enabled humanity's ascent throughout history in our quest of reaching for the stars. To have such lofty dreams, to actually dare to reach for the stars—we could not do so without understanding how our universe operates. So that we may learn to manipulate our surroundings and produce a world that we desire to occupy. A world of plenty. A world of security.

Through education, we can uplift the minds of our youth so that they may be capable of finding a better path forward in the future for problems that may arise—both as matters of our own making, as well as events that arise from the machinations of life progressing without regard to our presence.

Mike Hobart

How Did We Get Here?

Yeonmi Park discussed her story with Jordan Peterson about how she found her way from the depths of the North Korean nation all the way to the shores of the United States of America. She would then go on to attend Cambridge University, and encountered a level of resistance to free thought that was alarming. So much so that she came to the conclusion that her time and resources spent at the university were a waste.

What is the point of supporting educational institutions when the social environment is disincentivizing their own teachers and students from partaking in free speech and free thought?

How did we get here?

How did the western world, which has been propped-up as a beacon of freedom and opportunity, wind up in a position where our institutions are actively silencing narratives and positions that challenge popular thought?

How can a People actually hope to develop new technologies and new solutions to problems young and old without equally new thought?

You cannot get a new solution without a new approach, or new way of thought. Those new thoughts require opposition in opinions. Proper resolution is found within the conflict of ideas. Battles of the mind, body, and spirit present us with the answers to our greatest quandaries.

I have found myself revisiting one book in particular many times in the past few years: "*Tailspin*" by Steven Brill. The highway of our past that gives us the answers we seek is extremely nuanced and clouded with government legislation, civil battles, political strife and shifting cultural norms. I invite you to follow along on a journey that meanders through topics that will likely bore you to tears, but need to be acknowledged and reviewed, so we can do our best to avoid similar scenarios in the future.

Moats & Ivory Towers

In order to find the answer(s), or at least provide possible justifications for how it is we got in this predicament, we first look to the '60s.

Our first person of interest (POI) goes by the name of Russell Inslee Clark, Jr., former dean of admissions at Yale and the person responsible for revolutionizing Ivy League admissions strategy.

After the end of WWII, America's best and brightest returned home from the war, and in the decades following, had established a minor aristocracy among the upper echelons of socioeconomic hierarchies. Men and women hardened by conflict tend to be effective businessmen and leaders. As conflict may not deter them as easily as it would civilians. This establishment of a new aristocracy was also made possible by developments in standardized aptitude tests that colleges began to require in the late 1930s. Clark intended on disrupting this dynamic.

What Clark did at Yale was to cease taking admissions from wealthy underachievers, and instead he sought to give the most capable and hardworking from those with impoverished status a chance at something greater. At the time was arguably a great thing—Clark seemed to be evening out the playing field of academia within the wealth inequality gap. However, what he did not foresee was how those very intelligent boys and girls would position themselves later.

Following are a number of quotes from a Yale Law School graduation speech given by Daniel Markovitz in 2015, which most accurately detail what Clark's meritocracy metastasized into, "where he claimed that in earning a degree from 'the country's most selective law firm... actually marked their entry into a newly entrenched aristocracy that had been snuffing out the American Dream for almost everyone else,'" Brill wrote, citing Markovitz's speech.

"Elite lawyers' real incomes…have roughly tripled in the past half-century, which is more than ten times the rate of income growth experienced by the median American…

> "But it is perhaps... more surprising still to learn that the top 1% of earners, and indeed even the top one-tenth of 1%, today owe fully four-fifths of their total income to labor. That is unprecedented in all of human history: American meritocracy has created a state of affairs in which the richest person out of every thousand overwhelmingly works for a living."
>
> —Daniel Markovitz

In short, the new aristocracy was those who worked the hardest and smartest, not those who inherited the most.

"Elite lawyers' incomes will place you comfortably above the economic dividing line that comprehensively separates the rich from the rest in an increasingly unequal America," Brill wrote.

So, what does this mean exactly? Well, consider for a moment how much harder you defend that which you've worked hard to earn, versus defending that which was given to you. If you've sacrificed precious time, sweat, and blood to earn a position in the clouds, I would imagine you would not kindly give it up. You would entrench your elevated position. You would defend it. Not just for your own gain, but for your offspring, your family, and those you hold dear.

Markovitz continued...

> "This structure, whatever its virtues, also imposes enormous costs. Most obviously, it is a catastrophe for our broader society—for the many (the nearly 99%) who are excluded from the increasingly narrow elite.
>
> "Brewster and others embraced meritocracy self-consciously in order to defeat hereditary privilege, but although it was once the engine of American social mobility, meritocracy today blocks equality of opportunity. The student bodies at elite colleges once again skew massively towards wealth.
>
> "The excess educational investment over and above what middle-class families can provide that children born into a typical one-percenter household receive is equivalent, economically, to a traditional inheritance of between $5 [million] and $10 million per child. Exceptional cases always exist... but in general, children from poor or even middle-class households cannot possibly compete... with people who have imbibed this massive, sustained, planned, and

practiced investment, from birth or even in the womb. And workers with ordinary training cannot possibly compete... with super-skilled workers possessed of the remarkable training that places like Yale Law School provide."

Keep in mind, those numbers are from the speech in 2015. That is one decade and $8+ trillion in currency debasement (inflation) ago. If we assume a conservative 30% inflation rate since then, that would be $6.5 million and $13 million respectively.

Markovitz began to truly lay out how these entrenched positions come about, and how the disruptors end up becoming the incumbents. They become the very system that they sought out to destabilize.

"American meritocracy has thus become precisely what it was invented to combat... a mechanism for the dynastic transmission of wealth and privilege across generations. Meritocracy now constitutes a modern-day aristocracy, one might even say, purpose-built for a world in which the greatest source of wealth is not land or factories but human capital, the free labor of skilled workers."

—Daniel Markovitz

Today, that 'free labor of skilled workers' can also be reimagined as the data provided by our activities on social media platforms and interactions on the web. Data. Provided freely, 24 hours a day, 365 days per year.

Markovitz quite effectively laid out just how the 2015 Yale Law graduates had taken their first steps into a lifestyle amongst the American Elite that were effectively working to undermine efforts of the lesser classes to enter their ivory-laden existence. However, this isn't the whole story.

Eroding Our Future

In the United States of America, public school systems are maintained by way of taxes. This includes a combination of income taxes, federal and state taxes, property taxes and some fees sprinkled into the mix. With the vast majority of tax funding coming from state, county, and city levels. Public school systems are funded by the governmental body. Let's look at the path by which we got here.

Our journey begins in 1883 with the Civil Service Reform Act (aka, the Pendleton Act). What this act did was require standardized competency tests for federal civil servants, allowing hiring to be based on merit. Which I think we could all agree would be a positive. Let's look at how this system has been progressively gamed since its conception.

Fast forward. A man by the name of Charles Reich wrote an essay in 1964 that was titled "The New Property." In his essay, Reich argued that benefits and programs provided by holding a civil servant employment position should be treated equally as private property[3]. While only an essay, it appears that this suggestion was taken to heart by the bureaucratic government body.

Then we pop over to 1978 with the Carter presidency. Here we get the introductions of:
(1) the Office of Personnel Management (OPM)
(2) the Merit Systems Protection Board (MSPB)

These two offices tacked on additional rules to the hire/fire process of civil servants. What was added was a category designed to attempt to identify the best employees and to improve the efficacy of performance reviews, with the intent of encouraging the high-quality workers to seek positions more "deserving" of their work ethics (labeled "senior executive service"). Again, I (and I'm sure many of you as well) would argue that this sounds like a general positive, which is likely how it gained public support.

What this implies is that if you're employed by the government, your benefits (and by association; your employment) shouldn't be subject to termination/removal without proper justification. Again, this remarkably sounds like a benefit, right? This sounds like a level of protection for employees from any possible "good ol' boys club" dynamics from within government employment leadership.

---

[3] https://openyls.law.yale.edu/handle/20.500.13051/14924?show=full

The Second Renaissance

We have the establishment of the beginnings of a meritocratic system with the Pendleton Act of 1883, with civil servant employment being recommended to be treated as private property by Reich in 1964, and the meritocratic hiring processes being further strengthened by the OPM and MSPB in 1978, under President Carter.

The relationships of these mechanisms continued to get gamed within the judicial system culminating in a case in 2016 in Arizona. A woman by the name of Sharon Helman was appealing a case in which she was charged with a failure to practice proper oversight in her position managing the VA hospital in Phoenix, Arizona in 2014, where the waiting list lines and phony record-keeping under her watch were the foundation for a scandal. This scandal involved phony records in those very performance reviews established by the MSPB and OPM. The reviews showed exemplary service provided by the administrative staff, yet the data reflecting waitlist times and falsified record-keeping suggested otherwise. The outcome of Helman's appeal concluded that the charge brought against Helman did not accurately specify what kind of oversight she was expected to perform in her position as manager of the VA Hospital of Phoenix.

Sure sounds a lot like the early innings of DEI.

So, when an individual gets removed from a government position, they are legally allowed to appeal this movement—in the intent of protecting those who may have been fired without proper justification[4]. When an individual appeals their termination from a civil servant position, there's this neat little system where the state government foots the bill (partially dependent upon the outcome of the case).

But it isn't just footing the bill for the court fees. State governments also pay the appellant their salary while waiting for their hearing(s), as well as outcome, as Brill noted. This is thanks to a piece of legislation, also passed under the Carter Administration; the Civil

---

[4] https://www.findlaw.com/employment/wages-and-benefits/how-does-due-process-protect-a-public-employee.html

Service Reforms Act of 1978. With assistance from the Back Pay Act of 1966 awarding the appellant back pay (including interest and benefits) to cover the duration of their wrongful termination. And the Equal Access to Justice Act where the appellant may recover attorney fees and 'other costs' related to the appeal.

In a 2009 article published with *The New Yorker*, titled *"The Rubber Room,"* also published by Brill, describes his discovery of "rubber rooms." These rooms were facilities where teachers deemed incapable by New York City's Department of Education were housed during normal work hours. New York's civil service laws and union contracts kept all teachers awaiting arbitration hearings on the state's dime. According to Brill, these teachers would remain on the state's payroll for anywhere from two years to upwards of five while they waited for the backlog of arbitration hearings to clear.

The combination of MSPB and the OPM made for a quite-powerful mix when weaponized by the teachers union(s).

"[The] United States spends more per capita on public K-12 education than most other developed countries, yet its student achievement scores rank well below the countries long considered America's rivals," Brill pointed out.

Unfortunately, this system could be leveraged in urban and metropolitan areas of the country. The reason being that unions tend to be more popular, and have significant political pull, in areas that have low voter turnout and are dominated by democratic candidates.

In these environments public employee unions are incentivized to provide significant support for left-leaning party candidates that will, in turn, be incentivized to provide political (and financial) support to the unions that assisted in their election. This allows for unions to strategically position politicians that would be inclined to support their actions—like the appointment of arbitrators to appellant cases regarding termination of government employees over "implied" incompetence or "poor" performance.

Let's recap again.

We have unions leveraging due process and government employment (which is being treated as private property), justified by the MSPB and OPM, and rooted in the Pendleton Act, all of which is being paid for by the taxpayer. These appellant cases largely resulted in settlements or reinstatements, rarely ever terminations. Meaning the vast majority of the court costs and salaries were footed by the state, which directly involves the taxpayer.

Who Pays The Greatest Price?

I've laid out this complex series of mechanisms for a few reasons. I outlined how the gamification of this system is costing the taxpayers insurmountable amounts of money. But there are even more grave implications between the lines here.

These teachers that are being initially terminated, (let's assume for argument's sake that their termination was justified). Whom are being reinstated based more on legal technicality and less on merit, are going back and "teaching" our youth. Particularly in urban and metropolitan areas where poverty and poor education are very real and proliferate issues.

When we allow these individuals to return to these environments without recourse being taken against these failed teachers, how likely do you believe it is that they change their ways? How likely are they to put effort towards actually improving their capability? There's nowhere along this process that incentivizes these individuals to actually work towards being better teachers. They just got terminated, paid to chill, and reinstated, all while maintaining their pay and benefits. There's no punishment involved. When there's no punishment for poor decision making or incorrect action, nature justifies this as a rule stating, 'there is no need to improve.' Nature's correcting forces are pain, failure, and in its most extreme, death.

How does this affect the students? Do the students benefit? No. Do the parents (the taxpayers) benefit? No.

In fact, the community begins to deteriorate over time. The longer such a problem goes uncorrected and unchallenged, the more momentum it builds.

The low-quality teachers, the politicians, the unions, and the arbitrators all benefit. All while the children not only pay the cost of incompetence by way of being forced to tolerate impotent teachers, but they grow up in a reality where they lack the education to have any competitive edge in a university, let alone the job market. Worse yet, they inherit a country's debt burden (which has no chance of slowing down).

On top of that, when our public schools are directly funded by way of taxes, the government itself plays a very strong role in what is being taught within our schools. He who wields the money printer carries the big stick.

The past 20 years have been a pretty strong indication of who controls the printer: the Federal Reserve. And who can influence the Federal Reserve? Congress, the White House, and even more importantly now than ever, the market (equities, foreign exchange, and bond markets).

When taken into consideration, there are a lot of hands that can play a role in determining what is taught by public schools. But the real question is: Are they the right hands? Do we honestly believe that politicians, bureaucrats, lawyers, and bankers know how to teach best? Or what is best for the students of a socioeconomic status that they haven't personally been immersed in for decades, if ever?

What is one symptom of bureaucracy? The stifling of innovation, and the resistance to new ways of thought. Mix that with the unlimited printing capabilities of the Federal Reserve and you get a powder keg of compounding powers to not only resist change but perpetuate a significantly flawed and failing system, at the cost of each successive generation.

Our country and our leaders continue to sacrifice our children, and their futures, upon the altar of a falsified reality where America is still considered the "greatest in the world." Real leaders wouldn't offer up the futures of their children on a silver platter to their friends for the sake of holding on to the comforts of their positions.

We should be sacrificing our immediate comforts to build a foundation for which a brighter tomorrow can be built — if not to be enjoyed by our children, then built by them. That is the whole point to doing anything of value with the short lives we are given.

"Nemo vir est qui mundum non reddat meliorem."

Mike Hobart

CHAPTER 3: The Social Contract

America has been pushed from the pinnacle of learning, opportunity, and modern achievement into a corner of broad deception and decay. How did we get here?

Well, let us consider the benefits—and consequences—of generations developing and proliferating across a period of history after America establishing a new world order with the IMF, BIS and World Bank following the Bretton Woods Agreement. The ease of life that followed in the "peacetime" started (in greatest proportion) at the peak of the power dynamic —politicians and bankers—thanks to global reserve status, and over the decades, has bled down through the socioeconomic hierarchies, across corporations, brick-and-mortar businesses, and eventually down to the average citizen. Until the 1970s.

This easy lifestyle, enjoying the benefits of a position of leverage and power (thanks to the credit-based economy exacerbated by Nixon), has allowed complacency to flourish to catastrophic levels.

Ultimately, America's political and economic woes can be seen quite easily by looking at the general public's health crises. Plural. Physical, mental, financial, philosophical, and of course spiritual.

We have been receiving the signals for decades, yet we continue to overlook and even ignore these warnings and notifications. By ignoring these omens, we risk following the very same path as great empires of old. If we do not act to re-route America's health crises, which encompasses both mental and physical health, then I fear that America's time in the sun will end up resembling that of Icarus. These signals are easily identifiable within our health industries, which include — but are not limited to — hospitals and medical services, insurance and hospitality, fitness, food producers, diet and nutrition, and of course the pharmaceutical industry.

> *"The proper role of government, however, is that of partner with the farmer — never his master. By every possible means we must develop and promote that partnership — to the end that agriculture may continue to be a sound, enduring foundation for our economy and that farm living may be a profitable and satisfying experience."*
>
> — President Dwight D. Eisenhower, special message to the Congress on Agriculture, September 1, 1956

Civilizations And Health

David Montgomery is not a household name, but I believe that will change over the coming years and decades. As America continues down a path of economic hardship, thanks to the buckling of our Keynesian economic system due to such liberal monetary policy and protectionist, socialist measures of propping up nonproductive corporate and governmental ventures ... trouble is brewing just outside the gates.

As I've burrowed further and further down the investing strategies rabbit hole over the past years, I've learned much. What started with investing strategies and economics has led to further investigations in my first field of expertise; health, and then the money behind much of the "health sector," and recently to where our health comes from. The basis of all health would be nutrition — yes — but following first principles thinking, we come to the real genesis — farming and its bedfellows; ranching and animal husbandry.

The relationship between agriculture and the health of a civilization, especially in the death throes of empires, echoes across the ages. Montgomery covers this briefly in his publication, *Growing a Revolution*.

While the inflation angle is talked about religiously, pointing to the "why" a particular empire imploded upon itself, the food is rarely talked about as a final straw. I'm talking about **thee** straw. The one that broke the camel's back.

A societal collapse of historical importance can ultimately be boiled down to one very simple (but priceless) variable that any community requires: consistent food provisioning. This was Montgomery's revelation. Whether we're talking about the first communities that had begun cultivating crops some 12,000 years ago, or we're discussing active revolution from under a ruling power, the driver(s) of revolt ends up disrupting the food supply. Rome and Paris both had this pressure point in common: bread shortages sparked public outrage. But what ultimately led to these shortages? Was it greed by the farmers? Could they have been extorting prices of crops due to a monetary landscape that was producing hardship? Or was it, perhaps, gluttony by the citizenry? Driven by an insatiable hunger of an entitled populace that grew up during an age of plenty and relative comfort.

"Farming looks mighty easy when your plow is a pencil, and you're a thousand miles from the corn field."

— President Dwight D. Eisenhower

Earth and Empire

It was neither of these factors acting alone. What Montgomery describes on multiple occasions throughout his book is a series of events in history that don't technically repeat, but they certainly do rhyme. Where an empire reaches peak power, actual growth stagnates. The ruling class attempts to push growth further by manufacturing (fabricating) it through inflationary monetary policies that eventually bleed into aggressive and destructive agricultural and harvesting practices. Today, one of these fabricated growth strategies is the financialization of products and services. In the search for yield, these societies end up turning to strategies that sacrifice soil longevity and health. Eventually reaching a tipping point, where the soil can no longer support the desired output. These observations have even been echoed by great philosophers of antiquity, like Plato, Xenophon[5], and Aristotle[6] who identified this relationship of agrarian techniques and rapid topsoil erosion leading to hardships in their respective times.

This is a problem that has persisted since the first nomads decided to settle and sow what would become the seeds of farming strategies that would last across the ages. And we're still doing it, just with greater technology and some improvements on understanding. Yet, we're still making similar big mistakes.

With the addition of centuries of rapid technological innovation, this can also mean that we are driving the issue faster, and *farther*.

As I've been discussing and learning about the economics that our modern society machinates, the very obvious connection that I began to ponder was the effects of modern economic and monetary policies upon public health. I often found myself asking, "If our economy is so great, and we're 'the greatest country in the world,' why is it that we have some of the worst education ratings and health demographics on the planet?"

This was ultimately rage-inducing.

Montgomery goes on to write about the 2015 Agricultural Organizational Report, where it was stated that *"soil degradation erodes global crop capacity by 0.05% per year."* On top of this, a third of the world's agricultural land has been eroded. Going even further, despite heavy agro-chemical use by Western nations, up to 40% of global crop yields are lost to pests and disease. And then, down the line after said crops have been collected, ~25% of what is not lost to pests or disease gets wasted between production and consumption. Meaning that when it's all said and done, we're only capturing ~45% of the food we're producing.

---

[5] https://publicism.info/environment/dirt/4.html
[6] https://www.todayifoundout.com/index.php/2015/07/eureka-the-discovery-of-photosynthesis/

Mike Hobart

> *"Most people fail to realize that the lack of nourishment in our food today comes from the farming practices draining our soil's richness. Let's take oranges, for example. Today, we need to eat eight oranges to get our grandparents' same nutrients from one. Nutrient degradation in our soil poses a severe threat to our ability to feed 9 billion people by the year 2050 in a way that promotes vitality and a healthy way of life."*
>
> — Harry Gray, *We Are Overfed and Undernourished*

Addictions

The fertilizer industry is fascinating. Did you know that plants require nitrogen to grow, yet they cannot utilize the nitrogen gas that dominates our air? Nitrogen gas is extremely stable. Farmers needed a way to get nitrogen into a form that their crops could utilize.

How did we do it? Bombs.

…not *that* way….

During WWI, two scientists, by the names of Carl Bosch and Fritz Haber, devised a method to produce ammonia (NH^3) from nitrogen gas while searching for a way to produce bombs that delivered a larger payload. After WWII ended, bomb factories swiftly repurposed their ammonia expertise, giving rise to a new industry: fertilizer production. Decades later—some 50 or 60 years down the line—this shift had paved the way for giants like Koch Fertilizer and Cargill to dominate the market, as highlighted by Montgomery.

Montgomery cites Guy Swanson, a regenerative farmer, who compares the fertilizer industry to that of narcotics. It's a chilling analogy. Dealers offer a taste—an appetizer of quick yields—that hooks farmers fast. Yet over time, the soil weakens, yields falter, and the cycle deepens. Addicted, farmers return for another fix, chasing relief from the very problem the product creates. The Hegelian Dialectic.

The Second Renaissance

This cycle of dependency doesn't just trap farmers—it also wreaks havoc on the environment. Half stays. Half vanishes. According to Montgomery's work, only about 50% of nitrogen and phosphorus fertilizers get taken up by soil. The rest gets carried away in the water table, ending up in our rivers, streams and oceans. Right next to the soil that gets eroded away by tillage. My own home state of Iowa has seen 50% of her native topsoil carried away by this process since the days of the pioneers, with 30–40% of what is lost in recent history coming from gullies alone.

Soil and fertilizer might not spark lively debate over cocktails—I get it. But here's the thing: this matters. Ignoring it won't fix the mess we're in, and identifying the problem is step one.

So let's dig deeper. Why do we fertilize soil in the first place? And why can't we break this addiction?

"Burn down your cities and leave our farms, and your cities will spring up again as if by magic; but destroy our farms and the grass will grow in the streets of every city in the country."

— William Jennings Bryan

The Problem

We add fertilizer (including herbicides and pesticides) to our soils in order to drive the growth of our cover crops so that we can capture the greatest crop yield and get one step further in progress toward 'solving the hunger problem' — or so we are sold. But this still begs the question, why do we **need** fertilizer in the first place?

Let's pretend, for conversation's sake, that we have a farm with absolutely perfect soil quality and climate. We may not need fertilizers for quite a while, but if we're practicing modern farming techniques (aka, "extraction farming"), the soil will ultimately be degraded (because the nutrients have to come from the soil, and if those nutrients aren't being replenished at an equitable rate then we get...) to the point of needing nutrient supplementation via fertilizers. The reason is a multifaceted flywheel.

First, the process of tilling breaks up the surface tension of the soil, destroying root systems. Opening up the layers beneath to the extreme conditions of the open air. These conditions can come in the form of intense rains that can carry away this newly exposed soil, or it can come in the form of direct sunlight and heat, which dries out the soil. But the heat doesn't just dry the soil up, this heat also sterilizes the layers that have been newly exposed. This was emphasized by Rattan Lal (p.79, Montgomery), soil science PhD, and then confirmed when comparing temperatures of tilled farm soil temperatures to those of natural forests (a 20 degrees difference). Following the hot and dried up soil, tilled fields require more watering, but once soil temperature climbs above ~90 degrees Fahrenheit biological activity ceases. Meaning earthworms, root systems, and the microscopic organisms that make up the mycorrhizal biome that produces the glucose that plant roots feed upon. This is where we finally arrive at the point.

How often do we think that our farmlands encounter summer days where the soil temperature approaches 90 degrees? I assure you that I do not have the answer, but growing up in Iowa ... blistering hot days were quite common. When biological activity in soil stops, imagine it's like stopping an entire economy for two weeks and thinking everything will be copasetic after.

As if that could ever work...

But it doesn't stop there. Tilling already wreaks all sorts of havoc on soil health, and then we pile on another layer of trouble: herbicides and pesticides, dumped by the ton onto our fields. Even after engineering genetically modified crops (intended to be resistant to pests), pesticide use increased by 7% according to a 2012 study. Tilling tears at the soil. Erosion spikes. Biological activity—gone. Soil health fades, and with it, the nutrient punch of our crops. Then come the herbicides and pesticides, seeping into our water, poisoning us and everything we eat. All of which is producing a negative feedback loop on the longevity and strength of the soil upon which we base our entire lives.

One of the very important dynamics of this agrarian system that is also heavily overlooked, (or rather, in my opinion, willfully ignored) is the relationship that livestock and ranching play in this very complex ecosystem. Movements like The Beef Initiative represent a return to classic, regenerative methods of understanding the source of our food and the access to high-quality beef.

Mike Hobart

> *"Standard feedlot cows are fed genetically modified (GMO) soy and corn which contain toxic herbicides and pesticides. Cattle also receive hormone therapy to assist in weight gain, which allows the cow to be sold at a faster rate. Unfortunately, at times, profit is a priority over quality when it comes to supply and demand. This is why it is so important to weaponize your health by obtaining high-quality beef from an organic or regenerative rancher. Today's farmers and ranchers are strong examples of true conservationists. They have a deep love and appreciation for the land because it in turn supports their families."*
>
> — Texas Slim

Those animals also provide very important nutrients to the soil microbiome, of substantial note being nitrates and carbon. Nitrogen is that ever-important element that we mentioned earlier that plants need but many cannot pull from the atmosphere. They can pull carbon, luckily. However, carbon is a compound of high demand for life to flourish. There's a lot of conversation going on around pulling carbon from the atmosphere (which is not a great idea by the way), and very little being said around needing to get it back into our soil.

Without having a healthy and diverse ecosystem beneath the surface, our crops don't grow as strong and healthy as we need them to, so that our children can grow strong and healthy, so they can grow up to be smarter and stronger than ourselves, in order to solve the bigger and badder problems that they will face long after we are gone.

Livestock can heal the soil, but the real game-changer lies in how we farm. It's time to ditch the plow and work with nature's rhythm.

> *"A nation that destroys its soil, destroys itself."*
>
> — Franklin D. Roosevelt

We must farm in a manner that is much more aligned with natural mechanisms. Not by trying to force Mother Nature via synthetic compounds, herbicides and pesticides.

The Second Renaissance

You don't work against Mother Nature; she wins every time. You either work with her or fail.

The reason it's difficult? The strategy requires a complex scheduling of rotating cover crops between corn, soybean and wheat, as well as incorporating livestock rotations to provide manure for the fields. Very importantly, there is no tilling. By cutting out tilling, you maintain the surface tension of the root systems which act like skin by protecting the microorganisms that dwell just below the surface of our planet. This ecosystem is extremely sensitive to exposure to open air, temperature, weather and direct sunlight exposure. And even more important for the health of crops and soil diversity is that this wildly impactful mechanism provides very important nutrients to these plants that we need to nourish ourselves.

If we want to answer why our populations are so unhealthy and weak, we must look at why our foods are so unhealthy and weak. And if we are being honest with ourselves—which, right now we are having a very, very hard time with being honest with ourselves — the next step along the path leads us to the health of our soil.

No-till fights back. Crop yields climb. Fertilizer costs drop. Fuel bills shrink. Water stays put.

> *"Beck increased his soybean yields by 25 percent, from 63 bushels an acre to 79 bushels an acre. At the same time, corn yields increased as well, rising from 203 bushels an acre for continuous corn to 217 bushels an acre for a corn-soybean-rotation, and 235 bushels an acre for the more complex rotation, The whole system became more productive under a diversified rotation. And because it uses fewer inputs – less diesel, fertilizer, and herbicide – it's even more profitable (p.106, Growing A Revolution)."*

— Dwayne Beck, Director of Dakota Lakes Research Farm

On Beck's research farm they take all matters into consideration with their cover crop rotation studies, even so far as tire pressure — so as to avoid squeezing the water out of the soil, like a sponge. Oh, and with regards to that last claim of "requiring fewer inputs," Beck also gives us this bit of data:

Mike Hobart

> *"An increase of organic matter content from 1 percent to 3 percent can double the soil's water-holding capacity, while helping to prevent water logging that leads to the anaerobic conditions that favor soil-dwelling pathogens (p.98)."*

Dan Forgey, manager of Cronin Farms (20,000 acres comprising 40% cropped farmland and 60% native prairie pasture) has been incorporating a mixture of 10 different crops, he's been working to improve organic matter and soil carbon on his lands:

> *"So far they've increased their soil carbon by 1 percent. That may not sound like a lot, but [Dan] Forgey says that each percent of organic matter holds about $600 worth of nutrients an acre (p.109)."*

Now we have to answer, "why do farmers choose to till then, if no-till is so much better?" Government subsidies. If you've never met a generational, Midwestern farmer, these folks can be the most compassionate and hard working individuals on the planet (and lively to boot). Our farmers have been constantly under pressure for the past 50 years. As was infamously put by Earl Butz in 1973 to *"get big, or get out,"* America's Nixon administration made it quite clear that their desire was for America's farmers to industrialize (and centralize) in the search for fast money and big yield. This can hardly be disputed when considering the USDA secretary's words here in combination with Nixon's cessation of the gold standard. Since then the farmer's flywheel has been spinning faster and faster, allowing large farmers to rapidly expand and industrialize while their smaller neighbors are forced to either double down and grip the wheel harder, or risk being cast out. All the while, we are destroying generational livelihoods and our own country's health in the same breath.

Big farms won; soil lost.

> *"...cultivators of the earth are the most valuable citizens. They are the most vigorous, the most independent, the most virtuous, and they are tied to their country and wedded to its liberty and interests by the most lasting bands."*

— Thomas Jefferson

CHAPTER 4: Economics & The World Around Us

Lack of proper masculine role models and extended-order-effects of Keynesian economics are squeezing the broader male population and contributing to the rapid degradation of Western Society.

> "Observe, in this connection, the widespread phenomenon of men who are old by the time they are thirty. These are men who, having in effect concluded that they have "thought enough," drift on the diminishing momentum of their past effort – and wonder what happened to their fire and energy, and why they are dimly anxious, and why their existence seems so desolately impoverished, and why they feel themselves sinking into some nameless abyss – and never identify the fact that, in abandoning the will to think, one abandons the will to live."
>
> –Ayn Rand, *The Virtue of Selfishness*

Subtle Diffusion

One dynamic that I like to make a point of bringing-up in conversation is the mechanism of pressure transfer, or diffusion. Tensions in societies, nutrients in bodies, storms in the sky—one rule binds them. Energy flows. Molecules hum with it, vibrating between bonds. Add heat—like a pot on a stove—and they buzz faster. Bonds strain. Then snap. Water boils.

Societies work the same way. Communities, civilizations, even individuals hit boiling points when energy floods in unchecked—political heat, emotional strain, spiritual churn. That pressure's got to go somewhere.

It always does. Energy rushes from high saturation to low, a constant dance of flux. Nothing sits still. In a strong society it fuels purpose—a man channels restless drive into skills, drawing others in. But when it builds unchecked? Groups vibrate. Bonds fray—then shatter. Chaos scatters the pieces. Lone actors scramble for scraps, not thriving as a body anymore.

Scholars (should) see it—diffusion drives migrations, wars, and wealth. And it's shaking us now.

On Feminism

Rewind to the early 2000s. A generation teetered on the brink—smartphones, Facebook, and progressive waves pulling it apart. Feminism scored big wins, leveling the job field—bravo to those pioneers. But then the tide turned. Confidence swelled, goalposts shifted, and a slow swell against masculinity crashed over us.

Today, men drown in it. Two camps of cast-outs have emerged: simps and incels. Simps grovel, showering unreturned (and unjustified) pedestalization, sometimes extending to full blown deification. Incels stew, blaming the world—women, men, anyone but themselves. What relates these two archetypes together? No masculine guides. No sherpas to show them 'the way.' No one to show them how it's done.

The fallout's brutal. Attractive traits—sexual or otherwise—fade from men's toolkit. Peers, friends, bosses all crave them, but they're now scarce. Men naturally gather around other men of similar skill sets, backgrounds, belief systems, etc. Where a hierarchy within these communities will naturally develop. These decentralized male groups provide direction and guidance in young and adult male development. Now we've seen decades of degradation of these male communities and infiltration by women seeking acceptance into male dominated spaces. Interrupting the uniformity of the space (males leading males) and turning the environment into one of competition—over the female(s). What used to be learning and teaching environments have been transformed into watering holes. All thanks to the invasion of women into the worlds of men.

These cultural shifts impact the prevalence or scarcity of desirable male traits. Over the course of successive decades. What follows is a simple supply and demand marketplace environment. Then, modern society kicks things into overdrive. Social media distorts supply and demand perceptions. Online dating amplifies the male and female sides of the equation—where the vast majority of men have been proven to be who pays the cost.

Remember that Keynesian fiat hamster-wheel? It grinds them down. Financial hardship and pressures like the acute currency debasement of 2020 (let alone from the 1970s) keep individuals in prolonged, chronically stressful environments. Then throw in the cyclical market deviations like recessions and fears of losing one's job. Or the more recent outcome: young white males who cannot find jobs, thanks to the DEI policies and initiatives. Stress and trauma signal the release of cortisol. Cortisol is a hormone that triggers catabolism, which is a process our bodies use to break down organic material to yield adenosine triphosphate (ATP), the body's energy resource. Economics don't just shape wallets here—they mold spines, spirits, and bodies.

From the '90s to the 20-teens, we let these flaws fester. What was left behind is a brittle mess—too weak for truth, too proud to fix.

Psychology's Power Over Physiology

There is some evidence that men and women react differently to stress—shocker—as well as differently to varying forms of stress—who's surprised?

A study by Kirschbaum, Wüst, and Hellhammer[7] explored the differences in response to stress-inducing scenarios, the researchers found that men reacted with a 3-3.5x elevated cortisol (the "stress hormone") response to public speaking while women saw an increase of merely 1.5-2.5x. While these differences may not appear to be large at face value, it is vital for the reader to consider how exponentials can incite radically different outcomes with minimal differences in initial values.

Within the body we do not need massive variance in inputs in order to get seismic variance in outputs. Interactions within the body are expressed many millions of times over and over, some resulting in a mechanism that acts much like compounding interest. Each cellular function and interaction will produce a waste product, these waste products can accumulate when the system is not allowed (or able) to adequately clear the build-up. Individuals who are not adequately hydrated, or sleeping well, can see impaired functioning of these systems. On the production side. On the "clean up" side. Such dynamics have impacts beyond workout recoveries, this will affect mood regulation and states of mind, like depressive rumination.

Now there's the dynamic of how stress plays a role with regards to feelings of self-efficacy or having control over our lives. A meta-analysis found consistencies with the theory that individuals of higher social status tended towards having feelings of greater control, and therefore having lower cortisol levels[8]. Whereas those of the opposing position tended to express lower feelings of control and greater levels of cortisol.

To belabor this point even further, one study looked at how individuals to whom criticism was given relating to how a past

---

[7]

https://journals.lww.com/bsam/abstract/1992/11000/consistent_sex_differences_in_cortisol_responses.4.aspx

[8]

https://discovery.ucl.ac.uk/id/eprint/10085954/1/ShermanMehta_StressCortisolHierarchy.pdf situation was handled reacted to a following event[9]. When an individual was given negative feedback prior to performing a task, this resulted in worse anticipatory stress regulation; leading to greater cortisol levels, and a more negative stress response. While the opposite occurred when praise was delivered over handling of a past event.

Now where things get murky is over the discussion on whether cortisol has a directly inhibitory relationship on testosterone levels (also known as the Dual Hormone Hypothesis). The reality is: it depends. It depends on the individual, the situation, and any number of static and acute variables. Between a massive meta-analysis on the dual hormone hypothesis[10] and a study involving Olympic weight-lifting competitors[11] this reality is, in my opinion, accurately described.

The real moral of this story is how interpretation of self-efficacy and social status can, and do, yield very real physiological effects. And how these mechanisms and responses can, and do, effect and steer our behavior and interpretations of the world. Then it is also important to consider how these relationships will affect individual development and current, as well as future decision making.

A Developmental Question

Fighting for feminine empowerment and equal opportunity was a start—but it didn't stop there. The push morphed into an offensive, with masculinity itself cast as the enemy, necessary or not. There was little care for consideration of how this could (would) affect the broader population of boys and young men that grew-up and matured during this time in history.

---

[9] https://www.sciencedirect.com/science/article/abs/pii/S0018506X18304860

[10] https://www.sciencedirect.com/science/article/pii/S0149763417306784

[11] https://www.termedia.pl/Journal/-78/pdf-32347-10?filename=2_00643%20Article.pdf

That shift hit hard—namely for the boys growing up under its shadow. Imagine boys and teens, surrounded. Family, friends, teachers, doctors, celebrities—everyone—calling masculinity 'toxic.' Or blaming men for the world's woes, no nuance, no specifics. What happens then? Doubt festers. Self-loathing creeps in.

Let's not ignore the detractor position suggesting that 'not everyone was saying that masculinity was toxic'—yes, you would be correct. But what happens to language, especially language that is experiencing a fad social phenomenon? It gets used ubiquitously. Liberally. 'Toxic masculinity' would get used in hyperbole to simply describe activities by certain gentlemen, or certain groups of gentlemen, that were deemed wrong or undesirable. This is meaningful.

What this mechanism resulted in was two-fold unhealthy for culture. (1) the overuse of the term or concept greatly eroded the value or impact that the original turn-of-phrase invited—the public began to become desensitized. (2) what happened in tandem with the overuse was it got used for situations that were outside the purview of the original concept. Resulting in a much wider distribution of boys and young men hearing that masculinity was associated with toxicity, which is inherently bad (for everyone). Generations were taught to despise their own skin—over a sex they did not choose—Women might identify with the sting that this invites—so why double down? Gender roles took millennia to forge. They worked—reproduction, parenting, society. All breaking down and dysregulating. All of this makes for a disrupted family formation mechanism in western developed nations. To make matters worse, they were also discriminated against in the employment market, multiple times over if they were male, white, and of Christian faith. Which is the majority distribution for the United States and Europe.

The Second Renaissance

Home & Parental Environment

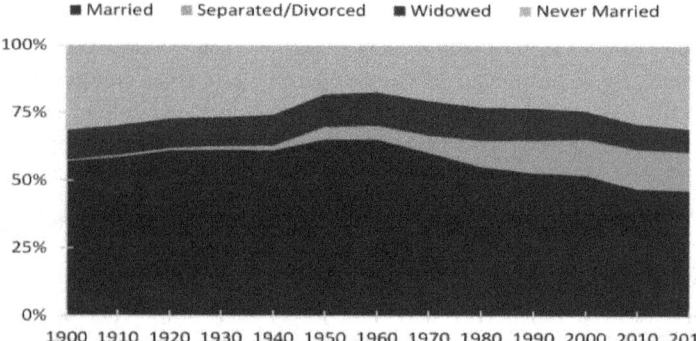

Figure 2. Current Marital Status of Women, 1900-2018

Source: NCFMR analyses of U.S. Census Bureau, Decennial Census, 1900-2010 (IPUMS); U.S. Census Bureau, American Community Survey, 2018 (IPUMS) Note: Data for separated were not available until 1950.

Figure 1 https://www.bgsu.edu/ncfmr/resources/data/family-profiles/schweizer-marriage-century-change-1900-2018-fp-20-21.html

Now let's ponder what kinds of effects can be yielded during a time when divorce rates were on a consistent trend up and to the right[12], while marriages have also been on a declining trend. What could possibly happen to the mental health and development of children being dragged through a fracturing of their nuclear family? Regardless of how volatile—or tranquil—a divorce may be, I assure you that no child wants to see Mom and Dad part ways. How could these children have their points of view altered, or shattered, by going through this kind of event, while hearing (and witnessing) society on the outside not only demonizing masculinity, but picking up momentum in this offensive?

---

[12] https://ourworldindata.org/marriages-and-divorces#divorce-rates-increased-after-1970-in-recent-decades-the-trends-very-much-differ-between-countries

Hollywood, cable television, music, literature, standardized education, university...how many of these were permeated with anti-masculine/anti-'patriarchy' narratives?

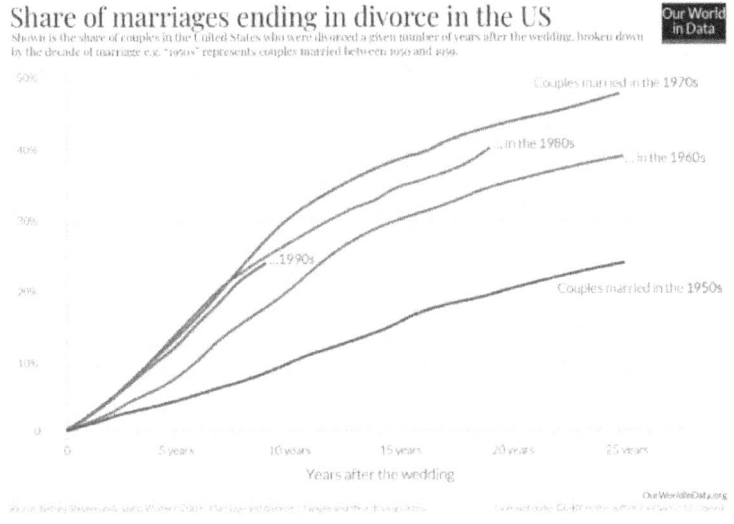

Figure 2 https://ourworldindata.org/marriages-and-divorces#divorce-rates-increased-after-1970-in-recent-decades-the-trends-very-much-differ-between-countries

Consider how hearing a particular way of thought, repeated ad nauseam can, and does, impact an individual's psychology or view of the world. Or consider how much such a mechanism can erode, if not destroy, an individual's psyche entirely. Not unlike information collected from the MKULTRA experiments[13]. Where the objective was to learn the manners in which human psychology & behavior can be steered, manipulated, or destroyed and remade entirely.

The handling of a parental separation will also have the potential to leave long-lasting effects on development in youth. Particularly how these children will view the opposite sex and the world broader. How many of us had grown-up witnessing negative interactions

---

13

https://digitalcommons.cedarville.edu/cgi/viewcontent.cgi?article=1005&context=history_capstones

between men & women, and therefore had our perceptions of the opposite sex take a more confrontational position?

"All men want is…"

"Women just…"

How many of these individuals have grown up approaching communication as a competitive environment? Where each discussion is an arena where the aim is to "win." This point-of-view can result in approaching personal relationships with a type of scorekeeping. Where instead of approaching each conversation as an opportunity to grow for either (or all) parties involved, it devolves into a competition. Wherein this environment further devolves into an attempt for domination, not just between men & women, but between factions on both sides of these lines. Resulting in an environment where everybody loses.

Fans of the movie "Arrival" will recognize this problem from a scene where Dr. Banks (played by Amy Adams) identifies this very same misstep where the Chinese attempt to use the game Go as a form of communication.

I'm confident that we've all fallen prey to these mechanisms. I for one fell victim to the 'He-Man Woman Haters Club' for an embarrassingly long time in my youth. Choosing to essentially show no regard for the humanity of the women in my company. They were only there to use me, and my friends, for temporary gratification or to gain access to one of our friends that they viewed as more 'alpha.' Which happened to all of us in college far more often than any female will ever admit to. This Merkle Tree of decision-making was very common for young men of my generation, and I'm very confident that our failings in maturity undoubtedly contributed to not only the militant transformation of the matriarchal movement but also added to the gravity-well of attraction for so many women (young and old) that have fallen victim to this system. Which only served to perpetuate this cannibalistic and parasitic relationship between American Men & Women.

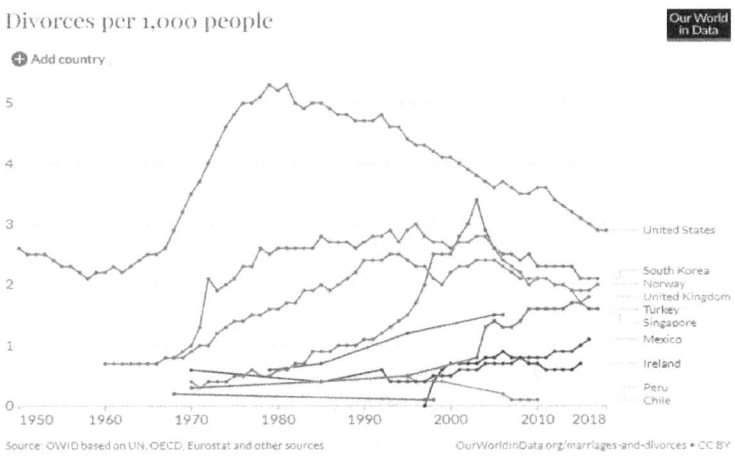

Figure 3 https://ourworldindata.org/marriages-and-divorces#divorce-rates-increased-after-1970-in-recent-decades-the-trends-very-much-differ-between-countries

Fortunately, the most important part of problem-solving is first identifying the problem, and then acknowledging that fault lies, in large part, with the individual and how we react to a particular situation. We may not have control over much of our external world, but we do have near-complete control over how we *react* to our external world.

Masculine Compression via Economic Asphyxiation

We have briefly peaked at how a mindstate such as stress can elicit effects upon physiology, throwing difficulty into mental health maintenance. There is also a drawback to our current economic and monetary system that is producing a mechanism that I do not believe many, if any, are looking into…

The problem with an economic system that is reliant upon credit (debt) is that the system ends up fabricating scarcity (hat-tip to Robert Breedlove for bringing this dynamic to my attention) due to the money being anything but. When a fiat currency is the basis for economic activity and civil organization, those of whom find themselves close to the money fabrication facility gain outsized power for the accumulation of assets and resources than their peers (and competitors).

Admittedly, this does not sound quite as nefarious as I suggest, on the surface. With some first principles elaboration you'll pick up what I'm putting down.

A currency that is not based on value, that is capable of being produced flippantly, is also at the whims of very human character traits. A currency, or a money (there is a difference[14]), is merely an economic tool that allows for greater social organization by allowing for broader economic activity beyond that of barter. When there is no limit to manipulation of the supply of such a tool, beyond matters of honor or morality, it is merely a matter of time before variables such as circumstance and greed erode the power that adhering to a code of conduct can provide. This also results in erosion of adherence to reality. With such power you can effectively avoid repercussions, for anything. For decisions high in risk. And those low in intelligence. Stupidity bears little consequence when losses can merely be printed away.

An Honorless Money

When the economic tool by which society organizes itself lacks adherence to rules (or reality for that matter), this very same lifestyle bleeds across the top echelons of socioeconomic status. Those at the 'top' of such a society are destined to, enmoat themselves with the power and social responsibility of controlling the fabled money printer. Suggesting that abuses of this power are not only inevitable but destined to spread further and broader across the population of this class of society. Bleeding into lower classes over time. As we are hardwired to learn from those we deem superior. Our children do it. We do it. Meaning that while the top rungs of the socioeconomic ladder enmoat themselves, those on the next rung down will begin to adopt the same tactic.

---

[14] https://www.wallstreetmojo.com/money-vs-currency/

The hilarity here is that if Ronald Reagan had understood this dynamic with foresight, he would've seen that Trickle Down Economics simply transfers the philosophy and culture of the social elite down to the rest of the populace; people will act in accordance with their incentives.

> *"Imitation is the sincerest form of flattery that mediocrity can pay to greatness."*
>
> —Charles Caleb Colton

In this system fraud, theft, and conspiracy become standards of business and activity. Leading to more and more average citizens recognizing that there is little-to-no consequence for illicit activity. Even worse yet, the average citizen begins to recognize that in order to "make it" one is most incentivized to lie and backstab in order to get financially ahead in life. Resulting in a society that begins to cannibalize itself. Coworkers are less incentivized to cooperate and grow as a team, and neighbors develop an "us versus them" state of mind.

This ultimately ends up producing a cancer within society that can snowball, ultimately leading towards implosion. A snowball which we have been experiencing.

The Divine Masculine

The truly masculine members of society are men of humble, confident strength. Possessing strength in body, strength in mind, and particularly strength in community, philosophy, spirituality and emotionality. Men that work together to provide, not only for their own but for those that have been permitted in their proximity. A community can accomplish far more than a single individual or family. Where a way of life provides effective service to family as well as thy neighbor. To effectively coordinate and organize men must interact with integrity and honesty; the opposite eats away at the soul of a community and individual. Resulting in necrosis. In divine masculinity failures are shouldered as equally as victories—at face value. A debt-based economic system is purely antithetical to this philosophy. An economic tool that bears no cost to produce bears no consequence for wrong-doing and failure.

The dollar of today promotes lies, fraud, and theft.

In this system liars, fraudsters, and thieves are the most capable. Where the honorable and strong members of society face out-of-balance difficulty to both success as well as proliferation.

As if life wasn't difficult enough. Might as well make things even easier for the scammers, schemers, and backstabbers—amiright?

This disincentivizes men and women away from developing honorable, desirable character qualities. Producing a positive feedback loop of rot and corruption all the way down to the formation of the family.

Now that we've explored the philosophical aspects of social organization requiring quality masculinity, let's look at a tangentially related mechanism.

As a fiat system manufactures scarcity because the economic tool of exchange is anything but scarce, resources of legitimate scarcity become more scarce for the broader population while becoming less scarce for those of higher social hierarchies. Whether this refers to gold, bitcoin, equities, food, water, or real estate—all resources become easier to acquire the closer one comes to the money printer.

In a debt-based system, where the average male consistently loses purchasing power and is incentivized against becoming a proper masculine figure, he is therefore less incentivized to aim for proper family formation. This is what we explore next.

The Divine Imperative

The male imperative is to provide; provide resources, provide shelter, and provide security.

In a society where the currency with which economic activity is consistently debased, each male's biological imperative is made more difficult with each passing cycle. In a society where the average male has been priced-out of being capable of acquiring land atop of which one can begin building a home, said male encounters further disincentivization to follow the imperative set before him by nature. In a society where legal and judicial dynamics within the marital system favor the mother—in custody, alimony, and child support—said male is further disincentivized to follow the mandate pressed upon him by his DNA. This male is incentivized into avoiding copulation for fear of child-bearance. Reproduction becomes a net-loss for many males.

This has numerous of its own downstream effects. More constricted access to mates for the average male as women are geared to seek out men that have security in resources and habitat. This also results in more constricted access to quality males for our women, as more and more men are pushed by circumstance to become a weaker man than he may desire to become—adopting behaviors like simping or acting out of dishonor to acquire income.

The average male cannot responsibly afford real estate today, does not make enough for subsistence (let alone ascension), and cannot financially survive a divorce or unplanned pregnancy. The average male is disincentivized from seeking-out sexual partners.

This also results in avoidance of male-male competition. A very natural and healthy aspect of male development and maturation. A natural part of mate selection is encountering competition for a desirable female. Either during the process of finding a mate, or defending a mate from suitors. This means that these males are not experiencing the full gambit of male-male interaction. As many males can attest to, some of our greatest friendships start with aggressive confrontation and head-butting.

A Detour Into Cringe

A man's qualities are developed and reinforced via his tribe of male counterparts; brothers-in-arms. Each member of these groups will help each individual grow while holding each man accountable. This is why men are naturally driven to form groups, or 'packs' (the cringe of this is so palpable, but it is unavoidable in its relevance). This is also how alphas are meted out naturally. He who is determined to be most capable tends to also become the leader by which the community turns toward for guidance and decision-making, whether by his own will or that of the group.

This avoidance of seeking-out of sexual partners, and avoidance of general male-male competition, results in boys and men failing to discover where they fall on their sexual market value (as coined by Rollo Tomassi). Without this knowledge, a male fails to aptly assess his likelihood for success when finding a mate. This therefore also effectively produces a limiting factor on the development of confidence and efficacy. What it also means is that the individual male fails to identify where he stands in society, and amongst the broader male cohort. Without this knowledge the average male does not know where to take himself. Say for example when attempting to begin on a journey of individual development and improvement. We can't know where we're going (let alone where we want to go) without effectively knowing where we are. Leveraged by knowing where we've been.

What this all suggests is that the path of least resistance for the average male is as follows:

- There's little justification to act honorably (doing so is already difficult prior to economic and social barriers).
- Leading to production of poisonous and traitorous social circles.
- Weak justification to find a sexual partner due to the social & financial risk that comes tethered with such decision-making (and made even worse with the proliferation of pornography).

All of which produce a society rampant with narcissistic-nihilists who have little-to-no interest in investing toward the future, ultimately preferring hedonistic lifestyles.

Their prospects are so bleak that there's little point in building for the future. Masking their lack of self-efficacy with pompous peacocking and maintaining thin skin with regards to challenge, or failure—thanks to lack of exposure to competition and being provided examples of proper male leadership. These fragile-ego'd individuals tend to project qualities they perceive of reflecting an "alpha" while seeking domination of all (or as many as possible) within their proximity—including friends, foes, and potential mates—and having highly volatile emotional swings.

> *"The chief task in life is simply this: to identify and separate matters so that I can say clearly to myself which are externals not under my control, and which have to do with the choice I actually control. Where then do I look for good and evil? Not to uncontrollable externals, but within myself to the choices that are my own..."*
>
> – Epictetus

Mountains of Debt & Doubt

We have a system that is disincentivizing proper masculine development, by incentivizing qualities that are parasitic and pathetic across the board. We have a system that is high in stress production. Not being able to be as 'good' as one may desire because the world is unfair, beyond the fairness born of nature, can be extremely disheartening, increasing stress. Lacking proper incentives and support systems to promote success, personal development, and family formation gives little to aspire towards; further increasing stress.

Then there's those that have fallen prey to any (if not all) of these dynamics and have found themselves buried under mountains of debt & doubt. These mountains have terrifying effects on confidence and efficacy, increasing the prevalence of cortisol within the body, meaning further stress.

Finding one's self in any of these scenarios, let alone more than one, if not all, will find difficulties abound. Motivation is likely at rock-bottom. Inspiration is a distant fantasy. Sleep is hard to come by (or over-indulged in). Appetite is either nonexistent or to-the-max. Anxiety tends to be either red-lining, or nonexistent as the individual has simply grown numb. Vices become salves to numbness and pain. Quickly becoming addictions as they are the only avenues that can provide immediate distraction—getting misperceived as relief.

None of this even considered the weight on public health enabled by a deficit of physical activity. America is particularly weighed down by a populace that is acutely anxious, obtusely overweight, depressed, poorly rested, and chronically malnourished. What is also not taken into consideration in our discussion here is the work done by Dr. Shanna Swan on endocrine disrupting chemicals (EDCs) and forever chemicals and their relations with our endocrine system.

Do you think any of the problems become worse, or better, under those additional filters?

Hint: None of it is made easier by incapacitating personal health.

Part 2

Value Exchange

CHAPTER 5: An Uncommon Look at Common Economies

I'm going to start this conversation with an approach coming from a health professional's point-of-view first, as the topic of economics tends to incur a sigh of frustration and boredom and is quickly followed by a glossing-over of the eyes. At the same time, I am going to be approaching the health angle from a position as a United States citizen, as that is not only my situation, but this discussion is particularly relevant to my neighbors in The Union — expressly those that continue to ignore the table-pounding by health professionals, from the doctorate level down to us plebeians.

However, that is not all that I intend to accomplish with these next few chapters. What I present to you is an attempt to connect the biological and economic worlds by addressing their shared realities. This requires some brief elaboration of complex mechanisms and broad generalizations, as these intricate systems involve seemingly endless levels of interconnection and nuance. As the topic of economics is consistently approached from primary and secondary vectors, I intend to take a more tertiary approach.

~~First a general breakdown. To establish a foundation with which to build our discussion. The current plight in the US is the pandemic that is chronic metabolic syndromes (let's refer to them as CMS's for short) that are gripping the population.~~

Mike Hobart

~~The human body is designed to move. More specifically to meet challenge. To face adversity. Your body is designed to be tested, rebuilt, strengthened, and then tested again. Incessantly and forever. At least until our bodies can no longer keep up. Looking closer at the operations of an organism you will likely notice something, arguably more marvelous.~~

CHAPTER 6: A New Way of Looking

The common view of a body is to look at it as one whole, hierarchical system. The brain acts as the control center, the skeleton being the frame, and the varying organs as support systems for the brain itself. I propose a "new" approach to looking at a body.

Your brain's an organ—same as your stomach, lungs, muscles, skin. They've all got their own functions, but they share a thread. The brain and spine handle the heavy lifting—signals, upkeep, housing the neurons and synapses that shape your thoughts, memories, decisions, moves. Muscles, meanwhile, tense or relax on cue from that nerve hub, even stepping up as a fuel reserve when the chips are down. All of them, every piece, built from the haul your gut delivers.

Bones catch flak—folks peg them as inorganic, distinct from muscle or brain. But they're not. They're alive—organic, growing, flexing to the weight of the world that you bear, just like your whole setup. That's the kicker: they're not so different. And that sameness runs deep, right down to the tiniest player.

Enter the mitochondrion. These little dynamos churn out energy—raw fuel for every cell's unique role. Sound familiar? It should. Bones, brain, muscles—they all hustle with what they've got, just like those mitochondria firing on all cylinders.

Zoom out. Look in the mirror. You're a cog in something bigger yourself: your job, your family, your town, your country. Each layer hums with purpose, powered by that same relentless energy.

We've now leveled the viewpoint of a few separate systems within our body; the brain, muscle, bones, and mitochondria, as all being more alike than they are different. My intent here is to destroy the hierarchical structure that many of us were raised viewing our bodies as. Consider all of them on an equal playing field, separate from the holistic bodily structure. By doing this we can view them like individual parts to a well-oiled machine. Specialized components, performing functions that act as cogs in a greater machine.

Now for what powers them.

Let There Be Light

"ATP is the fuel of life. It's an energy currency molecule — the most important source of chemical and mechanical energy in living systems..."

— Sunyoung Kim[15], Professor of Biochemistry & Molecular Biology | LSU

Life runs on a molecule: adenosine triphosphate—ATP for short. It's the fuel behind most biological motion. Sure, that's a simplification—energy transfer's a messy chemical dance[16]—but I won't drag you through the details.

All life requires energy. Our planet's primary source of energy is the Sun. In the form of photons (light energy), plant life here on Earth soaks up this energy and converts it into ATP with the help of chloroplasts, ADP (which is adenosine diphosphate, a precursor to ATP), and sugar — a process widely known as photosynthesis. This is the gateway by which light energy makes its entry into the ATP economy of our planet. This energy provides vitality for the plants themselves to grow and reproduce. Consider how much ATP is used in simply growing bigger (how much food do you eat in a day?). So when we look at the plant life that acts as our gateway of energy onto our planet, they are also containing large swaths of energy within themselves as they grow, as they reproduce and as they spread. In a way, you could say biological growth can be viewed as a form of energy storage, like a battery.

Checks & Balances

This is where herbivores come in. Herbivores feed solely upon plant life to supply them with the sustenance they require to also

---

[15] https://www.sciencedaily.com/releases/2010/03/100301091428.htm
[16] https://www.britannica.com/science/adenosine-triphosphate

grow and reproduce. Providing extension of the energy storage system. Our herbivores are then followed-up by the carnivores and omnivores, which gather and store energy by feeding on the herbivores and other living organisms. All of this amounts to distribution and accumulation of ATP.

Thanks to the quote by Professor Kim that was provided above, we get a step closer in the direction that I want to take this discussion. Before reading Kim's work I had (coincidentally) already been thinking of ATP as a currency of life. And now that she has assisted in bringing in the term "currency," let's explore that angle a bit.

The Economics of Life

If we look at ATP as the "currency of life," then we can look at an ecosystem as a mechanism for economic exchange of energy currency and accompanying resources. Trading between different communities, as well as individual actors. ATP provides a medium of exchange that enables processes beyond just the exchange of nutrients (resources) between cells and organisms — which resembles a barter system. Cells can signal for supply and demand in the body just like you and I can in our community.

Now consider how ecosystems in nature operate. They are truly quite fascinating.

Whether we look at the plains of the Serengeti, the mountains of the Andes, the rainforests of the Amazon, the Saharan Desert of Africa, the depths of the Pacific Ocean, or the frigid cold of the Arctic poles; life finds a way. Not just to exist, but to exist in balance. But it's not by way of not trying to perturb that balance. This balance is found as opposing forces collide with one another; as all aim to pass on their genetic code to the next generation as all are forced to contend over finite resources. Whilst defending themselves from those who seek the resources stored within their own bodies, or that of their offspring. In doing so, plants engage with predators (including insects, herbivores, parasites, disease, and other plants). Herbivores engage with predators, and predators & omnivores engage with everybody else (basically).

A balance is found.

As one population grows it provides a greater level of sustenance to another (what is referred to in the military as a "target-rich environment").

This provides a greater source of sustenance to another; the system charges on.

Providing avenues to eke out balance, as long as the natural order of checks & balances is maintained. This homeostasis can be perturbed if one population disrupts too many systems in such a way that risks the continuity of complementing actors.

Take the lionfish. It's not from the Southeastern U.S. coast, but it's tearing through it[17]. Built for the Indo-Pacific, this species thrived in balance back home—here in the Atlantic, it's overpowered. Now it's unleashing chaos along the Southern coastlines, a cascade of trouble.

When one species grabs too much sway, things tilt hard. That dominance doesn't just wreck the ecosystem—it can boomerang. As their dominance upsets the balance of ATP distribution, and the supporting systems that were in place can no longer be sustained, the dominator runs out of resources to support its own growth rate. Upsetting the ATP flow, shredding the support systems, and the big shot runs dry. Resources vanish. Austerity hits—famine, cannibalism, even extinction. The slate wipes clean, and the ATP economy rebuilds from scratch, if it can.

Competition & Progress

Amid all this, life pulls off another miracle through its brutal competitive environment. Randomness—DNA mutation—might just be its slickest move. Those tiny code tweaks churn out

---

[17] https://www.fisheries.noaa.gov/southeast/ecosystems/impacts-invasive-lionfish

differences in offspring. Subtle shifts that can make or break a species—boosting its odds with mates or sharpening its edge against predators and tough times. This is a feature of life, not a bug. This feature repeats, amplifying, with each successive generation.

> *"When compounded over a decade, gains are greater than losses because you keep building off of gains; whereas you experience losses and approach zero, future percent losses matter less in dollar terms."*
>
> — Ray Dalio, Principles For Dealing With The Changing World Order, p. 225

If we take Ray Dalio's quote from above and swap out a few of the words; "gains" and "dollar" for "progress" and "life" we get closer to my point. Life's moves don't tick at our pace. We feel compound interest over decades; life plays it out across millennia—way beyond our grasp. So, we will adjust for the timescale difference as well. When all of these substitutions are done, we are left with this passage:

> *"When compounded over a millennia, progresses are greater than losses because you keep building off of progresses; whereas you experience losses and approach zero, future percent losses matter less in life terms."*

Life's constant churn provides an efficiency mechanism that results in heavily specialized, powerful, and extraordinarily complex systems. While randomness may feel unfair…there is a method to the madness. As the universe doesn't desire to have energy shored up within aspects of life that don't play an effective role in the energy cycle, societies and markets are no different.

CHAPTER 7: Economies of Complex Systems

How Mother Nature chooses to find the most lucrative investments for energy is brutal, dynamic, and complex. As are markets. Do you think hedge funds and day traders care about whether the individual companies'—or their employees'—feelings are hurt? The answer is going to be a resounding "no." Their jobs are to accrue capital where their purchasing power will be treated best. This is precisely what Mother Nature does with her ATP allocations. ATP accumulates to the organisms that are best fit to survive and reproduce. So too do economies allocate purchasing power to companies (and individuals, i.e. Elon Musk) that are expected to be as equally fit to survive and prosper.

To gain an understanding of what determines the resilience & survivability of such entities I recommend reading Robert Kiyosaki's *Cashflow Quadrant* literature as well as consuming content from Preston Pysh on cashflow analysis.

We've already explored how natural selection produces the most effective organisms. The very same machinations provide the most effective companies and individuals in a society.

We turn our attention to the events of the 2008 Global Financial Crisis and the intervention of bailing-out entities that were deemed "too big to fail."

Companies that were facing a brutal death because of foolish decision-making, aversion to responsibility and accountability and general lack of critical thinking by leadership, were kept alive. American leaders, in their short-sightedness, felt that it was best to avoid the immediate pain of these organizations going up in flames (not to mention the embarrassment that it meant for America's brand across the globe). What has resulted is a coterie of large American organizations being placed on government-backed life support. These gluttons have now not only swollen to even greater size but have replicated.

The Economics of a Body

Let's unpack a 'calorie.' Most of you know the big C 'Calorie' from nutrition labels—not just a quirky cap, it's a kilocalorie, 1,000 little calories packed into one. I flag this because punctuation can shift the game, especially in science. Yep, that 2,500-Calorie daily target? It's 2.5 million calories.

Now, for a little nutritional information.

Food's Caloric punch varies. Some nutrients hit harder—bioavailability's the trick. Fats top the list at 9 Calories per gram, alcohol kicks in 7, while proteins and carbs both clock 4. But here's the rub: fats pack the biggest load, yet the oxidative system takes its sweet time; carbs bring less, but the glycolytic system churns out ATP fast.

Here's a quick tie-in for the econ-minded. Think of carb digestion as day traders—fast profits, high energy, constant hustle. Fats? They're macro traders, or investors—big returns over time, less grind. One's a sprint, the other's a slow burn.

So, we've got the Calorie basics down—how they yield from different sources. Back to ATP, life's currency. Whether it's the glycolytic path (carbs mostly) or the oxidative one (fats), the rule's simple: more Calories, more ATP. But here's the catch—sourcing matters. Efficiency drives an economy, right? Same goes here. We can't dodge quality's role. We've hit it once with zombie companies;

now meet zombie compounds—food's dark side, courtesy of human meddling.

Carbs, Fats, Calories & ATP

ATP powers ecosystems through exchange, just like currencies and cash drive economies. So, here's a question: what happens when that value—ATP or capital—gets faked?

Look at processed foods—artificial sweeteners too. It drives me nuts, and I'd bet the exercise science crowd agrees bioavailability matters, big time, yet it's ignored. Natural foods—fruits, veggies, meat—digest clean and fast. Millennia shaped us for that, with enzymes like amylase splitting sugars in your spit or protease ripping proteins apart. Processed junk's a clunkier fit—less bioavailable.

Our system's precise—fragile, even. Feed it unnatural configs, and enzymes fumble. The gut microbiome, that crew in your intestines, stumbles too. Call these 'false foods'—mimicking nature's flavors with manmade junk she never greenlit.

It's worth sizing up natural foods against processed ones. Natural stuff packs extras—fats, amino acids, fiber, water—teammates that boost absorption. That's why whole foods outshine processed junk, powders, and supplements every time.

Now, I can hear it gnawing at you—how's this tie to economics?

Inflammation in Biological Economies

Lately, the nutrition, physiology, and psychiatry crews—plus plenty others—have been mapping the messy ties in our complex, tough, yet fragile bodies. Ties we've mostly ignored, willful or not. It's not a stretch to buy that what we eat messes with our moods—

short-term jolts or long-haul shifts. Research backs it: man-made or natural, food tweaks the gut's bacterial balance[18] [19].

This slots right into our earlier chats on economies, competition, sourcing, and natural selection. Pump resources into one group, and it'll outgrow the rest—more numbers, more turf. That means more space hogged, and more waste piled up. So, what do humans ace?

> "...stress got put into a part of the system that should not absorb that much stress."
>
> — Dr. Andy Galpin with Dr. Andrew Huberman[20]

We churn out waste—tons of it. And it's a mess out there: plastics clogging the Pacific, fertilizer runoff from monocrops, climate gripes galore. Call it inflammatory, every bit. The point? Our societies and economies spew reactions—waste's fallout. Inside us, with the gut bacteria we roll with, it's the same deal.

When we keep eating these inflammatory foods—stuff that pumps up certain gut bacteria—their waste piles up right alongside their growing numbers. Hold up, that's not news. The body's always churning out byproducts; it's just how creation and breakdown roll. What's wild, though, is the downstream ripple—second, third, fourth-order effects. Psychotherapist Tori Rodriguez breaks it down: some compounds, via our microbiota, tweak moods, even spark disorders or chronic messes[21]. An additional study from 2013 backs it up[22].

We've got to quit playing dumb about this. Personal health, ecosystems, economies—doesn't matter. Tiny, steady interactions can snowball into massive outcomes, and we can't keep shrugging that off.

---

[18] https://pmc.ncbi.nlm.nih.gov/articles/PMC6363527/

[19] https://wellnessmama.com/health/artificial-sweeteners/

[20] https://youtu.be/IAnhFUUCq6c?si=Vr37WJ2JVevWa21H

[21] https://www.psychiatryadvisor.com/features/exploring-the-link-between-persistent-infection-inflammation-and-mood-disorders/

[22] https://pmc.ncbi.nlm.nih.gov/articles/PMC3772345/

Inflammation in Human Economies

If fake, man-made ATP sources spark waste-driven cascades—disorders, nasty side effects—in our bodies, why wouldn't the same hit an economy of businesses and customers when it comes to fake currency? We've been hashing this out: ATP exchange in life mirrors a town or nation's economy. Waste from one system feeds the next, rippling into consequences bigger than they look—slow shifts that need time and distance to spot.

History's littered with proof. Song Dynasty China[23], the Roman Empire[24], today's U.S. Petrodollar[25]—fake currency, fiat junk, stirs inflammation. It kicks off chains of chaos, bending behavior as the mess unfolds. Digest a currency thin on resources—or one losing its heft fast—and trouble brews. Same goes for food: scarf down nutrient-weak junk, laced with artificial sweeteners or chemical sludge, and digestion falters[26]. Without the right backup, waste piles up—stuff that'd clear out if the real deal were on the plate.

---

[23] https://en.wikipedia.org/wiki/Jiaozi_(currency)

[24] https://mises.org/mises-daily/inflation-and-fall-roman-empire

[25] https://bitcoinmagazine.com/culture/the-hidden-costs-of-the-petrodollar

[26] https://nutritionj.biomedcentral.com/articles/10.1186/1475-2891-9-51

Mike Hobart

CHAPTER 8: A Tangled Mess We Weave

Our modern economy, juiced by the internet, zaps messages—and now value—across space and time in a snap. But here's the snag we're stuck in: this money-moving machine's a mess, riddled with hiccups. The big glitch? It's not a sleek, internet-born setup. Nope, it's stacked on a creaky old financial system. Double-spending forces a maze of checks and reversals—middlemen juggling red tape across jurisdictions, regulators, and banks, all tripping over each other.

Picture this: cells and organs in your body playing middleman—vetting, reversing ATP swaps billions of times a second. Thought would crawl, reactions to the world's chaos would stall, and survival? Done—let alone thriving. Now layer in the economic inflammation we've been hashing out. It's a compounding mess, a system too choked to live.

Economies hooked on fake energy—ATP or capital—breed collapse. Fraudulent sources spark negative loops, inflamed by ditching the natural backup that resource-rich systems lean on. Think Darwin: real competition, dense energy, free markets—life and economy both thrive that way.

Our economies—local and global—and our own health in the U.S. are inflamed to the gills, flat-out sick. A currency that's geopolitically neutral and legit—packed with effort and energy—

might turn the tide. Not a tech rollback, but a leap: new tools built for real, resource-dense exchange.

Currency's half the fix, though. Competition's the other. Let it loose—test tech and tactics in the wild. Without that, ideas stay stuck in books, unproven. Reality's messier than the chalkboard says.

Here's my take: our global economy, and all the little ones feeding it, mirror our bodies and ecosystems—tough, fragile, and beat to hell by our own hands. From you and I to the big shots, we'd better clock these parallels before we tank ourselves.

"Nothing in the world is worth having or worth doing unless it means effort, pain, difficulty… I have never in my life envied a human being who led an easy life. I have envied a great many people who led difficult lives and led them well."

— Theodore Roosevelt

And yeah, there's more…

Mike Hobart

CHAPTER 9: Let's Talk Energy

The interplay of biology, energy, and economic systems underpins this discussion, necessitating an examination of fossil fuels' role within the broader macroenvironment. Before proceeding, it is imperative to address a prevalent critique of oil: the ecological impact of spills.

Oil spills undeniably inflict severe initial damage on ecosystems—a point beyond reasonable dispute. However, advances in bioremediation mitigate their permanence. This technique employs hydrocarbon-digesting microorganisms, their efficacy enhanced by supplemental fertilizers, to degrade oil residues.

"In 2001 and 2003 the National Oceanic and Atmospheric Administration (NOAA) conducted random sampling of 4982 pits dug at 114 sites in Prince William Sound to determine how much residual oil remained; these studies found that 97.8% of the pits had no oil or light oil residues even though these sites had been heavily to-moderately oiled in 1989."

— Biodegradation and Bioremediation: A Tale of the Two Worst Spills in U.S. History[27]

---

[27] https://pmc.ncbi.nlm.nih.gov/articles/PMC3155281/

Oil & Energy

With this clarification established, the focus shifts to humanity's broader relationship with energy. Across history, harnessing energy has directly correlated with societal flourishing.

Harnessing energy = elevated prosperity

This principle manifests in harnessing kinetic energy for tools and weapons, or combusting biomass (e.g., wood) to generate heat, denaturing proteins for digestion (cooking), and achieving sanitation. The true leap, however, was mastering oil as an energy source, laying the groundwork for unprecedented technological advancement. Much as consistent water supplies were prerequisites for public health improvements[28], reliable energy access—yielding steady electricity—was essential for technological progress. Scientific inquiry demands controlled conditions and replicable constants, both reliant on uninterrupted power.

Technological advancement extends beyond feats of engineering or military might; it elevates living standards through iterative refinement. Each iteration yields more efficient, cost-effective solutions, reducing the effort required to achieve elevated outcomes. This process democratizes access, enabling broader participation in technological benefits—a higher standard of living emerges not from singular breakthroughs but from sustained refinement, contingent on reliable electricity.

Fossil Fuels & Powering the World Economy of Value

The nexus between energy and innovation parallels that of energy and economic vitality. Economies depend on consistent energy access, albeit for distinct yet overlapping reasons. Operational reliability—e.g., a refrigerator's utility hinges on a stable grid—is

---

28

https://www.researchgate.net/publication/250142768_History_of_water_and_health_from_ancient_civilizations_to_modern_times foundational. Yet beyond functionality, energy fuels innovation, driving iterative improvements that produce superior goods at lower costs and accelerated paces. This enhances both societal living standards and market efficiency, progressively shrinking the intervals between advancements (Moore's Law).

If energy is a currency of progress, time surpasses it in value. Fossil fuels, by underpinning reliable electricity, remain integral to this dynamic, sustaining the economic and technological systems that define contemporary flourishing.

Part 3

The Second Renaissance

Mike Hobart

CHAPTER 10: Where Did the Time Go?

Prior to delving into the core argument, an examination of monetary theory's temporal implications is warranted. In the early 2020s, I authored an article exploring how Bitcoin reframes individual perceptions of time, challenging the precepts of Keynesian economics and Modern Monetary Theory (MMT). Proponents of MMT assert that uncirculated dollars—those not perpetually traversing the economy—represent wasted potential, a stance predicated on maximizing monetary velocity. This metric, inherent to all currencies, quantifies the frequency of exchange. The contention lies not in velocity's existence but in its exaltation as an end unto itself, wherein savings are systematically disincentivized.

MMT posits that individual frugality—maintaining a financial buffer to mitigate future instability—lacks societal value. Such a view implicitly devalues personal sovereignty and resilience, framing them as burdens rather than bulwarks against dependency. It denies the average citizen the prudence afforded to hedge funds and conglomerates, presuming that economic stability derives solely from relentless circulation rather than strategic restraint.

MMT advocates further contend that inflating currency—eroding its purchasing power—is essential to stimulate growth, effectively appropriating time from individuals. This approach presupposes that economic vitality hinges on pressuring citizens into expenditure, such as via an arbitrarily 'calculated' inflation rate that functions as a gradual expropriation[29]. Analogous to a "slow boil," this mechanism seeks to constrain agency, rendering escape improbable by design. Consequently, entrepreneurs—spanning diverse sectors—have grown reliant on cheap debt, leveraging perpetually optimistic forecasts to inflate valuations and normalize high-risk ventures. Such practices, entrenched by recency and availability biases, belie the serenity from which innovation can also emerge, underscoring a tension between coercion and liberty in economic thought.

---

[29] https://www.cfr.org/blog/history-and-future-federal-reserves-2-percent-target-rate-inflation-0

CHAPTER 11: Breaking The Paradigm

Why can't regular folks save in something—cash or an asset—that grows because everybody wants it, there's only so much to go around, and the market's nuts for it? And why's it wild to think we can handle a money like that without screwing it up? Sure, people mess up—greedy, vain, easy to trick. You bet. But haven't we seen plenty showing that the big shots with power lie, cheat, and swipe whatever they can when it suits them?

Ever wonder what happens when you stash your savings in something solid like that? Your worth climbs just by sitting tight with an asset that's got no human baggage—no CEO caught with their pants down (happens way too often), no boardroom clowns tanking the ship (yep, that too), no bank fumbling your cash away. It's simple, steady—a little win for keeping it real.

CHAPTER 12: A Spoonful of Liberty…

Here's what happens—I've lived it myself.

You get freedom. One less nagging worry in the daily grind.

We had that vibe in America after WW2. You know, when Europe was a wreck, Russia was tapped out, and the world leaned hard on our oil and gold stash—pumped up by Allied wins snagging loot across Europe and North Africa. On the gold standard, regular folks thrived, saving in something finite. Life was good.

I can hear the grumbles already. Truth is, not everyone was hoarding gold bars—most were stashing cash tied to it, pegged so it grew as gold climbed. For a solid 25+ years, G.I. Joe had it made. Then the bigwigs—Keynesians under Nixon—called it quits on that "freeloading" party.

Stashing your worth in something fools and fanatics can't touch? That's freedom, not control. Gold was the old-school hero, but it's fading fast—too clunky for today, nabbed by the state, useless for privacy in a world that's lost it. You've got to chase the edges now, where innovation lives. That's where you dodge the stress of hoping slick politicians—like champ Nancy Pelosi—won't tank your cash with insider trades while they're at it.

Mike Hobart

The future's still a mystery, sure. But with something like Bitcoin, you've got a network ticking along—updating the ledger every 9 to 10 minutes, double-checking it forever. No slip-ups, no crossed wires—reliable.

CHAPTER 13: Stress Relief

That's a big deal. It hits deep—body and soul—for anyone plugging into a system like that. Ditching those money worries frees up your headspace, your spirit, giving you room to chase new stuff. Maybe you tweak your health, shake off lifestyle ruts held back by cash stress. Or pick up a new skill, a fresh passion. Grow your family, start one from scratch? Launch a business with tricks you've learned in that open air? Maybe all of it?

We're wired to hunt down passions—hobbies that stretch us, spark creativity, but still feel good. Building a killer gazebo for summer nights, tending a thriving garden, tickling the piano keys— it's the same itch. We love getting good at what lights us up.

Here's the laugh: Keynesians and MMT'ers swear regular folks won't bother. They'll just loaf. Me? I've been there—downing booze at a scary pace—and trust me, it's not as common as we think. We see party pics and assume everyone's wild. Nope. Most folks barely have the cash or hours for that. Worse, both sides—partiers and grinders—end up stuck with junk food and no energy, trapped in the Fiat Hamster Wheel. It's a nasty loop—health tanks, and climbing out takes years.

Add today's healthcare mess—prices soaring because we're all dragging from bad choices—and it's a pile-on. Systems meant to help turn into budget-busting habits. We slog for cash, dreaming it'll buy us freedom someday—less grunt work, a car to roll where we want.

But dang, we're clever at tricking ourselves out of it. That's where the Hamster Wheel snags us.

So, here's a question: what'd you do if money stress dialed way back? Not a magic fix—life's not that cute—but enough to spot a path out, a plan that clicks. A goal worth chasing that doesn't grind you down like now. What would you do with that kind of time, that peace, that shot at a freer future?

CHAPTER 14: Free Time. Free Spirit.

What would you do if that big, gnawing worry—money stress, feeling less-than, or just a bleak take on your future—lifted?

Some would party hard, no doubt. Others would kick back, nap off years of missed sleep. Fair enough, who wouldn't? A few would jump straight into building stuff—companies, gadgets, fixes the market can't resist. But that's the wild ends of the curve. Most of us—90, 95%—land in the middle. So what does that look like?

It's a sweet spot, where freedom and passion tug us along. We're born to make things—art, clever workarounds, being a better parent, you name it. It's an itch we've got to scratch with effort. Boredom? Not our thing. We're wired to tweak our world, whatever lights our fire—gardens, gizmos, tunes.

I wrote a piece a while back—kind of out-there, real feely—that digs into this. Every passion's a gift to someone else. A pianist fills a bar with magic (my personal soft spot), a painter slaps up a mural, an engineer solves a shopkeeper's headache, a personal trainer shares know-how. We've all got something to toss on the table, even if we don't see it. That's the glue—how we spark each other to build better.

Mike Hobart

CHAPTER 15: Valued In Absence

Freedom's gold for a society. It lets your heart and head roam—new paths, fresh fixes, bold ideas. And it gets a big boost when money's cut loose from the state. We've tried some slick moves to split them over the years—millennia, even—but it always circles back to the same spot: power hoarded, control tightened. Then come the abuses, the blowback, the tear-down, and the rebuild, hoping for better.

Word is, the French once broke out guillotines for that gig.

Pull money away from the state, and you yank one root of that mess. It loosens the grip on folks—not a quick high, more like dodging bigger trouble down the road. Success here looks quiet: no disaster to spotlight proves it's working. It's insurance for us humans.

Most folks don't get it, and that's why weak men brew hard times. They chase the quick buzz—comfort, luxury, now—shrugging off prep and grit. Whims rule, resilience? Nah. But strong men? They build good times by skipping today's treats for tomorrow's steadiness. They grind now so next month, next year, life's lighter—for them and the neighbors too.

The short sighters miss it: tackling tomorrow's headaches today cuts future costs. It's a buffer—time, space, peace of mind—to handle whatever pops up next. Splitting money from the state? That's a tomorrow fix we've kicked down the road for decades here in the U.S. Left us stuck—average Joes and Janes squeezed by the Fiat

The Second Renaissance

Hamster Wheel and their own cravings. They don't see it: their spark's been nabbed, dulled, tamed.

Here's our shot to show we're not those weak guys. To prove we're strong ones—ready to pay today's price for a brighter tomorrow, not just for us, but for everybody.

Mike Hobart

CHAPTER 16: Revolutionary Momentum

Given the ubiquity of the internet age, familiarity with Moore's Law—positing the exponential doubling of computational power—may be presumed. Yet, a less-examined dimension warrants scrutiny: beyond enhancing efficiency and capacity in labor, technological progress amplifies cognition, emotion, spirituality, and creativity. By automating work, technology liberates cognitive resources, enabling the pursuit of novel problem domains. Herein lies the foundation of this inquiry—the genesis of a compounding societal shift.

Consider the Renaissance. Its significance—cultural rebirth, scientific awakening—may be recalled, but its temporal scope merits closer examination. Such transformations do not arise spontaneously; societal evolution parallels biological maturation, as from caterpillar to butterfly or child to adult, albeit on a grander scale and protracted timeline.

Across the 15th and 16th centuries, radical thought permeated society, gradually suffusing individual consciousness. Yet the intellectual, scientific, cultural, and artistic hallmarks of the Renaissance were seeded in the 13th and 14th centuries. This lag invites interrogation: why such duration? The answer resides in communication. The velocity, depth, and reach of communicative protocols dictate the pace of ideological dissemination. The printing press, introduced in the 15th century, catalyzed this shift—an accelerant igniting a volatile blend of political, cultural, spiritual, and scientific currents, centered in Italy. Prior to this innovation, handwritten correspondence prevailed, prolonging transmission and comprehension. Revelations required extended gestation; epiphanies unfolded over decades, constrained by delivery delays and contemplation intervals. Commerce, too, labored under these temporal burdens, as trade and monetary exchange demanded commensurate patience.

Consequently, capital mobility and systemic reorganization faced stringent temporal limits. Communication speed serves dual roles: a constraint on, and a catalyst for, transformative impact. Compounding this dynamic was the populace's baseline intellect and education. Conceptual complexity impedes rapid diffusion—an idea's spread hinges on its accessibility to its recipients, akin to a virus: slow infection yields slow propagation.

This principle underscores the efficacy of simplified explication, as epitomized by the contemporary heuristic, "Explain Like I'm Five" (ELI5).

Mike Hobart

CHAPTER 17: History's Lesson in Today's Flavor

The contrast between the Renaissance era and the present is stark. Communication and economic exchange now approach instantaneous velocities, with near-total global reach. Educational attainment among individuals in developed nations surpasses that of six centuries prior by orders of magnitude. Yet, this acceleration coexists with a pervasive temporal deficit: vast populations, exemplified by the average American, experience compression across material, mental, spiritual, and emotional domains, ensnared by the FHW.

Economic compression stifles creativity and passion—the lifeblood of original thought and critical inquiry. These faculties demand both the capacity to engage problems from diverse perspectives and the impetus to resolve them. Their erosion precipitates a withering of vitality, a phenomenon observable in recent decades. Technological innovation has faltered; dominant entities resort to imitation, yielding a stagnation in the quality and diversity of market offerings.

Amid this plateau, monetary systems present an exception. A nascent digital asset—epitomized by Bitcoin—offers unprecedented efficiency in value storage, transfer, and exchange. By excising redundant intermediaries, it promises to purge economies of systemic waste while simultaneously incentivizing low-cost energy generation and grid stabilization[30]. Its decentralized architecture resists manipulation by any single actor, be it individual, organization, or state. Upon achieving market saturation, this asset could supplant conventional metrics like GDP—rendered dubious when 40% of a currency's supply emerges within a 24-month span—as a more robust gauge of economic vitality.

This innovation portends a fundamental reorientation of individual behavior. Where profligate expenditure once prevailed, frugality gains traction. Where nihilistic outlooks obscured viable futures, a pathway emerges—modest yet actionable—requiring neither financial acumen nor speculative trading to safeguard wealth against inflationary erosion. Simplicity, a requisite for the average citizen, underpins this shift.

Low time preference, thus incentivized, fosters an environment conducive to restored health, healed psyches, and burgeoning inspiration. The nuclear family flourishes under such conditions, enabling effective parenting and disciplining adult immaturity rather than indulging it. Concurrently, wealth redistribution accompanies these societal realignments, channeling resources to novel actors with distinct perspectives. This facilitates the funding of innovative solutions—both to emergent challenges and persistent inefficiencies—previously stifled by recursive aversion to change.

The caveat remains: such transformations unfold gradually, constrained by the temporal scale of systemic evolution.

---

[30] https://simplybitcoin.substack.com/p/bitcoin-mining-power-games

Mike Hobart

CHAPTER 18: The Incipient Energy Revolution

The preceding discussion concluded with a reflection on time—a resource unequivocally finite, admitting neither regression nor acceleration beyond biological limits. Societal paradigm shifts, akin to a renaissance, entail rapid, transformative changes across thought, culture, arts, sciences, technology, and commerce. Such revolutions require protracted gestation to achieve critical mass—a volatile confluence awaiting ignition. Contemporary society stands at this precipice, poised for a catalytic event or technology. In the first Renaissance, the printing press served this role; today, I contend Bitcoin fulfills an analogous function, albeit through an unconventional mechanism: its symbiosis with energy systems.

Emergent Catalysts

Instantaneous communication, as delineated previously, spans the globe, facilitating equally swift monetary flows. Technologies such as additive manufacturing (e.g., 3D printing) democratize production, enabling individuals with requisite materials and connectivity to fabricate diverse goods. This convergence constitutes an unprecedented reservoir of potential—ideas exchanged freely, funded rapidly, and realized with historic alacrity. Yet, a critical constraint persists: energy.

The Second Renaissance

WE SOLVED NUCLEAR DECADES AGO [31]

Energy underpins all processes—biological, urban, and industrial. It binds molecular structures (e.g., water's hydrogen bonds) and sustains individual and societal vitality. In technological innovation, energy availability mitigates the friction of experimentation. Beyond the inherent challenges of iterative failure, the costs of raw materials and operational infrastructure impede progress. Affordable energy generation emerges as a linchpin solution.

---

[31] https://youtu.be/4aUODXeAM-k

WHY YOUR ELECTRIC BILL IS SO HIGH [32]

Energy extraction, however, encounters escalating inefficiencies. Depleted accessible reserves—such as oil—necessitate intensified efforts to secure deeper deposits, elevating resource demands across exploration, extraction, and processing. Concurrently, extant energy forms (e.g., oil) enable the infrastructure for exploiting denser alternatives (e.g., natural gas), which in turn support electricity generation, polymer synthesis, and advanced engineering. This interdependence fosters a positive feedback loop: oil fuels natural gas extraction, which powers technologies that enhance productivity and habitability, amplifying societal capacity.

Economic Constraints and Bitcoin's Role

Unrestrained energy production remains elusive, governed by economic principles of supply and demand. Energy producers, like all market actors, balance output against consumption patterns.

---

[32] https://youtu.be/fFXcBIFoxdE

Overproduction risks market destabilization and insolvency, threatening the continuity of electricity provision. Basic supply-demand dynamics illustrate this equilibrium:

Static Supply + Static Demand = Price Stability
Increased Supply + Static Demand = Price Decrease
Static Supply + Increased Demand = Price Increase
Decreased Supply + Static Demand = Price Increase
Static Supply + Decreased Demand = Price Decrease

Counterarguments positing infinite substitutability ("someone else will step in") falter under scrutiny. Expertise and reliability—akin to preferring a seasoned physician over an untested novice—preclude cavalier disruption. Producers thus calibrate expansion to gradual demand growth, a cautious cadence that curbs societal potential.

Bitcoin mining disrupts this stasis. Operating continuously, miners generate persistent demand, yet possess the flexibility to halt operations instantaneously without equipment degradation. This tertiary revenue stream incentivizes producers to scale generation beyond conventional thresholds. By decoupling supply expansion from societal lag, miners enable a recalibration:

Increased Supply* + Static Demand = Decreased Prices
(*Supply augmented by Bitcoin miners, adjustable to societal demand shifts)

This dynamic gradually reduces per-kilowatt-hour costs for consumers, sustaining producer viability while attracting residents and enterprises to regions leveraging affordable energy. Enhanced tax revenues and land value appreciation follow, reinforcing economic resilience.

Over time, diminished energy costs lower production expenses, though not instantaneously. Competitive pressures compel market actors to optimize efficiency or cede ground to rivals offering comparable value at reduced prices. This iterative adjustment, rooted in supply-demand interplay, amplifies technological and economic potential, catalyzing a renaissance-scale transformation.

Mike Hobart

Societal Transformation

How many nascent ideas and aspirations teeter on the brink of realization today? How many linger unspoken across global populations, stifled by prohibitive energy costs, resource scarcity, or barriers to collaboration? Contemporary society stands poised for a transformative surge of unprecedented magnitude.

Public discourse fixates on the specter of global conflict—World War III—yet the envisioned future lies beyond this apprehension. War remains contingent; change, however, is inevitable. Fear of transformation is not obligatory, yet its arrival is immutable, irrespective of individual assent.

Bitcoin mining exemplifies this catalyst. The symbiosis of its persistent power demand—characterized by reflexive adaptability—with the steady output of nuclear reactors offers a compelling empirical foundation. Miners have already substantiated their utility in flare gas mitigation and integration with hydroelectric, wind, and solar systems; nuclear energy now emerges as a viable frontier. This pairing—harnessing surplus and waste energy while stabilizing grid loads—promises to reconfigure the structural underpinnings of modern civilization.

CHAPTER 19: Reviving The American Dream

How would you characterize today's landscape of relationship building and family formation? In my circles, it's historically been met with unease—a quiet dread, not bold fear. Why might this be?

I contend that fiat-driven distortions permeate every corner of existence. We've already discussed how the landscape is egregiously impacting the male population. Collectively, we overspend—buying unneeded goods to dazzle strangers, crafting personas for likes and clout. It's a hollow chase. Finding a partner becomes an emotional and intellectual slog, so draining that many just opt out. Imagine: you're drawn to someone's curated life, only to find the shine's a sham. At its worst, the relationship's foundation rests on fiction—like a building on sand, doomed to crumble. Failure's baked in.

Why this masquerade? Why project false lives for attention? Multiple threads tangle here, but one looms largest: money. The average single person can't swing a desirable life and stack meaningful wealth. Rent, phone bills, car payments, health insurance—costs pile up while wages lag inflation's relentless climb[33]. Living outpaces earning, plain and simple.

---

[33] https://www.pewresearch.org/short-reads/2018/08/07/for-most-us-workers-real-wages-have-barely-budged-for-decades/

This friction throttles a core human drive: partnering up, raising kids. The fix? Fix the money. Restore power to the nuclear family. Family structure matters—not its shape or the parents' identities, but its stability. It's a bedrock for kids, sure, but also for society. Healthy families fuel progress—America's beacon of freedom depends on them charging toward prosperity.

I hold that everyone's got greatness in them—not a moral stamp, but raw potential. A stable home unlocks it, teaching basics like socializing while igniting curiosity—those flickers that grow into passions. Passions shape us—hobbies into careers, quiet joys in our heads. They can nudge history itself. Who knows what inventions we've missed because someone's spark got snuffed out?

Family setups vary—Bitcoin won't iron out every kink. But it crafts a monetary system where people can thrive. Saving in an asset immune to debasement builds a cushion—not just cash, but peace of mind. It flips the script: the future's not a void, but a reachable better day. That shift bolsters happy, lasting families. Money woes tank marriages—Gen X and Boomers know it too well.

Bitcoin yanks the American Dream from nihilism's muck, hoisting it toward dawn's light ("*...at dawn, look to the East*"). It's a horizon worth chasing, not a blank stare. When folks—or a whole society—lose faith in tomorrow, despair festers. Depression spikes, addiction soars—escape hatches from a dead-end now. A flourishing economy won't erase that for everyone, but it can dial it back. How many loved ones wrestle depression or drugs, trapped with no exit? How many delay kids under a cash crunch?

Bitcoin revives that post-WWII glow—families buzzing, communities tight, futures wide open. Its decentralized, untouchable nature makes the Dream global, not just American. It's optimism you can bank on—a richer, fairer, freer tomorrow. Hope worth holding.

Part 4

The Energy Renaissance

Mike Hobart

CHAPTER 20: Politicization Of Energy

"That Would Be the Road to Hell for America."

— Jamie Dimon[34]

In the United States and Europe, two factions within the energy generation sector—oil and gas versus renewables—have contended for dominance over recent decades. Advocates for renewables seek the phased elimination of fossil fuels, while opponents highlight the dependency of renewable infrastructure on fossil resources across production, deployment, operation, and maintenance cycles, paradoxically amplifying demand for the contested fuels. Despite substantial government subsidies supporting renewable expansion, debates persist over operational efficiency and economic viability.

The Environmental, Social, and Governance (ESG) framework, once rigorously enforced, has encountered significant retraction. European policy has reclassified natural gas and nuclear power as sustainable, reflecting pragmatic concessions. Major entities, including BlackRock, S&P Global, and Shell, have abandoned carbon credit strategies, paralleled by a broader retreat from ESG rhetoric in corporate spheres, as exemplified by McDonald's.

34

https://x.com/tomselliott/status/1572682957562978304?s=46&t=kKMXNIef1KBLWfgv3xXtGQ

Energy, a universal necessity, transcends ideological divides; its politicization belies the shared demand for affordability and abundance.

Reducing energy consumption undermines systemic integrity, accelerating degradation. Climate-related justifications, irrespective of their merits, obscure this dynamic. Empirical evidence—such as NASA's February 11, 2019, report documenting increased vegetation across Asia correlating with industrialization—challenges prevailing narratives. Lectures by Tom Gallagher (linked below) further contextualize Earth's climatic variability, underscoring the complexity of atmospheric and hydrological systems. While also providing evidence that challenges climate catastrophization narratives.

PALEOCLIMATOLOY PART 1 – TOM GALLAGHER [35]

Energy consumption cannot decelerate without compromising societal progress. Advancement, pursued across myriad domains concurrently, necessitates work, which in turn demands energy. Sustained innovation and efficiency gains—essential for forward momentum—require increased energy inputs to develop and

---

[35] https://youtu.be/K6tWEjkEiZU?si=AaRvdsq8v0XKD2wP

implement. Such progress may reduce energy allocation to routine tasks, redirecting resources toward novel methods that streamline previously labor-intensive processes. Alternatively, genuine technological or methodological breakthroughs can introduce entirely new modes of operation. Curtailing energy consumption undermines these dynamics, impeding the capacity for both incremental and transformative development.

PALEOCLIMATOLOGY PART 2 – TOM GALLAGHER [36]

Sustained growth in energy consumption is indispensable for societal functionality and incremental advancement. Analogous to biological systems, stagnation in energy utilization precipitates decline; absent continuous expansion and the overcoming of constraints, systemic degradation ensues. The absence of adaptive pressure fosters dysfunction, accelerating collapse along a compressed temporal trajectory.

---

[36] www.youtube.com/watch?v=iZSYSWPYEbU

PALEOCLIMATOLOGY PART 3 – TOM GALLAGHER [37]

The nexus of energy availability, consumption, and infrastructure underpins societal advancement, with power accessibility serving as a critical determinant. Subsequent analysis will address emergent technological developments poised to catalyze a substantive transformation within the energy sector.

---

[37] www.youtube.com/watch?v=YMHKt9ylPpQ

CHAPTER 21: Energy and Growth

Stern, Burke and Bruns in their 2017 analysis[38] concluded that access to electricity is not sufficient for economic growth but that electricity use, and GDP have a positive relationship. Simply providing the availability of a resource does not dictate advancement, it is the use that results in advancement. Common sense.

"As a result, energy is an essential factor of production and continuous supplies of energy are needed to maintain existing levels of economic activity as well as to grow and develop the economy (Stern, 1997). There may also be macroeconomic limits to substitution of other inputs for energy. The construction, operation, and maintenance of tools, machines, and factories require a flow of materials and energy. Similarly, the humans that direct manufactured capital consume energy and materials. Thus, producing more of the substitutes for energy requires more of the thing that it is supposed to substitute for. This again limits potential substitutability (Cleveland et al., 1984)."

—The Impact of Electricity on Economic Development: A Macroeconomic Perspective (2017)

A confluence of excessive Environmental, Social, and Governance (ESG) prioritization, vilification of oil and gas, and climate catastrophe narratives has precipitated a narrow focus on emissions within energy generation discourse. This emphasis

---

[38] https://escholarship.org/uc/item/7jb0015q

sidelines critical considerations: reliability of electricity provision, infrastructural capacity, and the limited substitutability of hydrocarbons given current technological constraints. Such discourse aims to impose restrictions on energy and power consumption—an inherently anti-growth stance. As established, curtailing growth within an ecosystem undermines resilience, inviting systemic collapse.

Prioritizing emissions over availability, reliability, capacity, and cost efficiency erodes existing infrastructure. This degradation impedes the development and deployment of innovations that enhance efficiency, including reductions in waste and pollution. Resultant inefficiencies in energy generation and electricity supply elevate production and living costs, compounding into diminished living standards and exacerbating systemic vulnerabilities.

The politicization and tribalization of energy initiatives—manifest in the denigration of hydrocarbons and nuclear power, the elevation of wind and solar, and the omission of hydropower—introduce pronounced fragility into established economies.

> *"While solar energy is abundant and inexhaustible, it is diffuse compared to fossil fuels, and plants only capture about 1% of the energy in sunlight. Therefore, the maximum energy supply in a biomass-dependent economy is low, as is the 'energy return on investment' for the human-directed energy expended to extract energy. This is why the shift to fossil fuels in the Industrial Revolution was so important in releasing constraints on energy supply and, therefore, on production and economic growth (Wrigley 2010).*
>
> *In spite of this, core mainstream economic growth models disregard energy or other resources (Aghion and Howitt, 2009), and energy does not feature strongly in research on economic development (Toman and Jemelkova, 2003)."*
>
> —*The Impact of Electricity on Economic Development: A Macroeconomic Perspective* (2017)

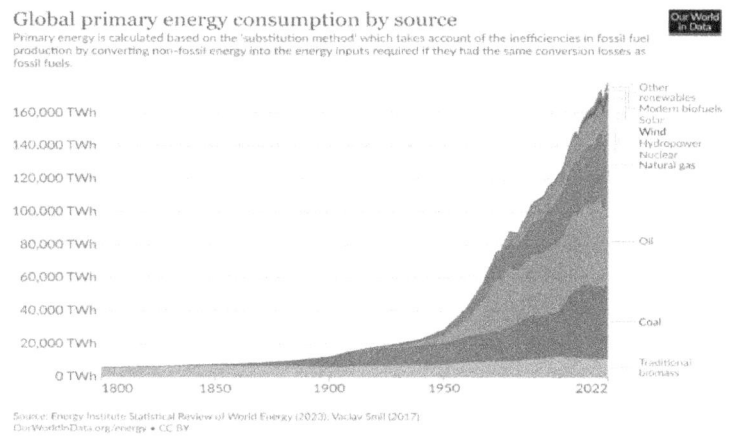

Figure 4 https://ourworldindata.org/grapher/global-primary-energy

Attempts to mandate the phase-out of established energy resources through legislative coercion, rather than market mechanisms, constitute an inefficient allocation of time and resources. Such interventions, as exemplified in legislative exchanges involving Jamie Dimon (referenced earlier), risk systemic disruption if enacted. Moreover, escalating power costs would likely provoke substantial resistance, potentially reinforcing the existing framework and rendering the effort futile irrespective of outcome.

A resilient society sustains entrenched energy sources while prioritizing efficient and reliable alternatives to enhance operational efficacy. This approach facilitates the concurrent development of innovative energy solutions, optimizing economic returns on generation investments. The resultant uplift in living standards establishes a reinforcing feedback cycle, advancing societal capacity.

Let's look at the investment relationship with regards to energy generation, capacity, and infrastructure itself.

Energy And Return On Investment

Stern and Kander (2012) determined that population growth absent a commensurate increase in energy supply precipitates a decline in economic output. Adapting the Solow Model, they integrated a low-substitutability energy resource (e.g., oil and gas) alongside labor into their projections, contending that prevailing economic frameworks undervalue energy's role in sustaining economic vitality, particularly in developed nations with robust power access. Their analysis further posits that expanding energy supply in tandem with population growth—leveraging technological enhancements in generation—elevates output. This augmentation, by increasing energy availability and utilization, amplifies GDP, a dynamic applicable even to advanced economies.

THE SOLOW MODEL AND THE STEADY STATE [39]

---

[39] https://youtu.be/LQR7rO-I96A

The Solow Model

The Law of Diminishing Returns, when modeled alongside constant depreciation and integrated with investment levels and their rates of return, yields a trajectory approaching equilibrium over time, eventually transitioning to negative returns on extended horizons. This dynamic bears significant relevance to energy infrastructure and output within the context of civil development and economic vitality. Underdeveloped nations adopting modern energy generation and distribution technologies experience substantial initial returns, which diminish as access and utilization saturate. Conversely, nations with ubiquitous reliable power—such as the United States and European states—encounter lower returns on investment relative to emerging economies leveraging contemporary strategies to close developmental gaps.

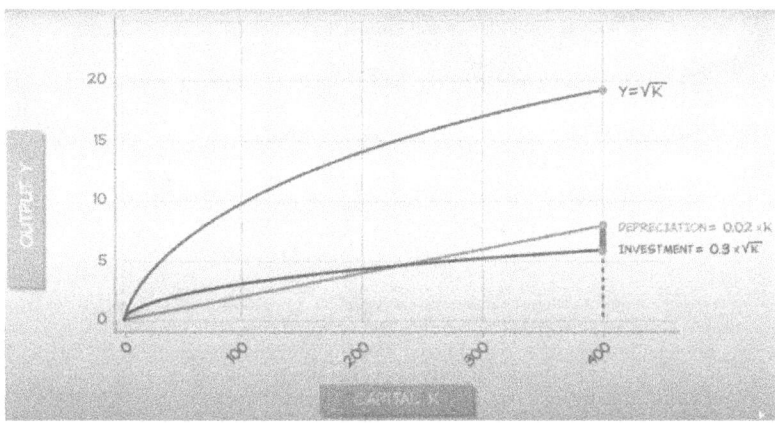

Source: The Solow Model and the Steady State, Marginal Revolution University

This analysis further indicates that inadequate adoption of advanced methodologies and technologies for energy production, capture, distribution, storage, and utilization incurs escalating maintenance costs that erode investment returns. Such inefficiencies divert increasing resources toward sustaining existing systems, merely mitigating degradation without advancing output. Continuous enhancement of energy-related capabilities is thus imperative; cessation of efforts to optimize sourcing, capture, distribution, and consumption precipitates systemic collapse.

The Solow Model & Energy

Stern and Kander demonstrate that integrating technological advancements in energy generation with a growing population and enhanced energy utilization extends the viability of the Solow Model, forestalling the break-even threshold. This dynamic sustains consistent GDP growth, as evidenced by the United States' economic trajectory over the past two centuries.

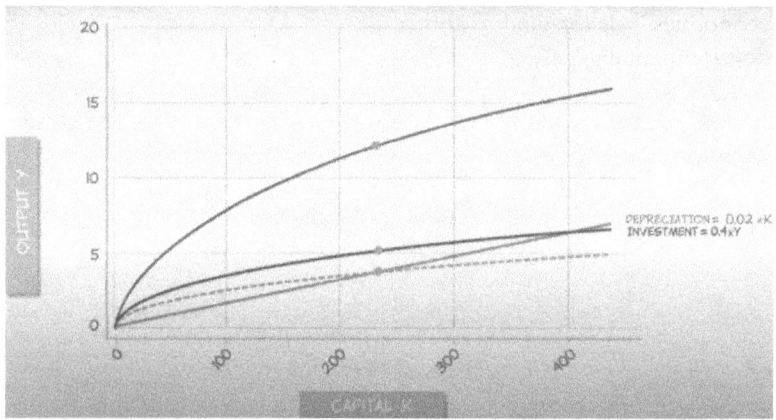

Source: *The Solow Model and the Steady State*, Marginal Revolution University

Advancements in energy generation through augmentative technologies and methodologies, coupled with expanded access, capacity, and population growth, drive the effective cost of energy toward zero. This trajectory reflects heightened returns on energy consumption, as increased work output accompanies rising utilization levels.

Figure 3: Sweden 1800–2000, Input Prices *(continued)*

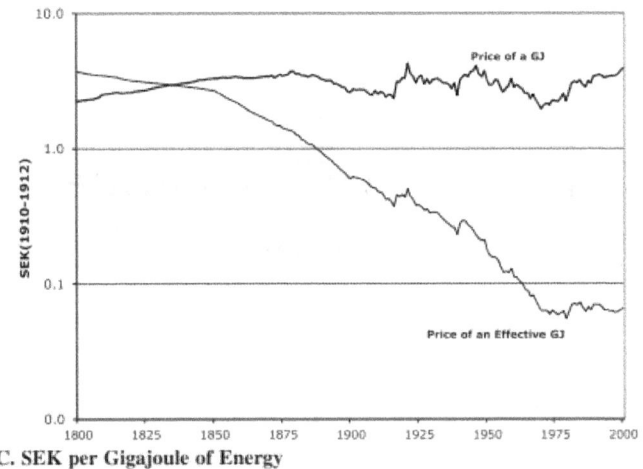

C. SEK per Gigajoule of Energy

Figure 3: Sweden 1800–2000, Input Prices

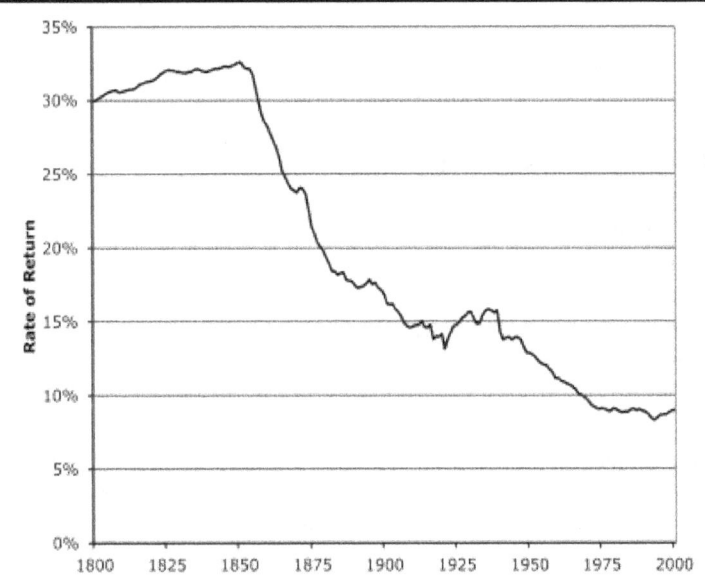

A. Rate of Return to Capital

Figure 5 https://web.archive.org/web/20240301023434/https://crawford.anu.edu.au/distribution/newsletter/research-newsletter/pdf/Energy-Journal-Stern.pdf

CHAPTER 22: The Future Of Energy: Bitcoin Mining

"Time-series analysis (Stern, 1993, 2000) shows that energy is needed in addition to capital and labor to explain the growth of GDP. But mainstream economics research has tended to downplay the importance of energy in economic growth. The principal models used to explain the growth process (e.g. Aghion and Howitt, 2009) do not include energy as a factor of production."

—*The Role of Energy in the Industrial Revolution and Modern Economic Growth*, Stern and Kander

Given energy's pivotal role across all economies, its marginalization in research and discourse merits scrutiny. Equally perplexing is the pronounced politicization and segmentation within the energy sector. Divisive narratives obfuscate a fundamental imperative: maximizing generation capacity without destabilizing economic systems, thereby sustaining societal functionality. Achieving this objective necessitates direct monetization of energy production.

A primary obstacle emerges: power demand exhibits significant volatility, fluctuating diurnally and seasonally. This instability extends to energy forms, particularly in economies subject to climatic variability or constrained by limited source diversity.

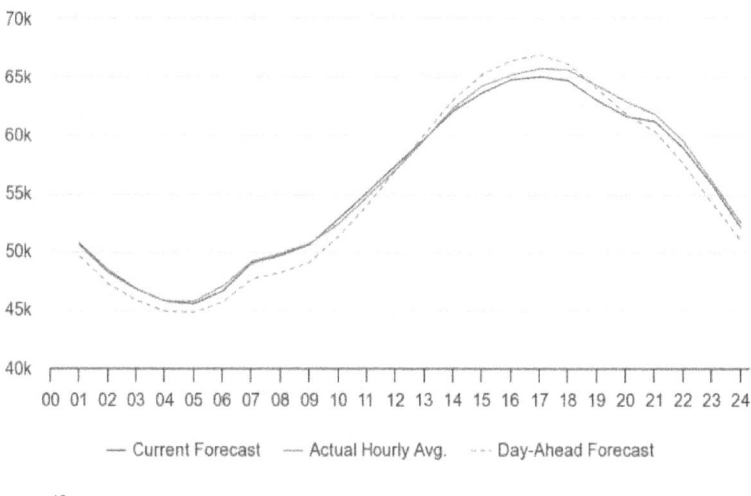

[40]

Is there a way for us to smooth-out this demand volatility so that energy producers can maintain a consistent run-rate while still being capable of providing reliable power to societal fluctuations?

The Future Of Energy

The feasibility of this approach is affirmed: Bitcoin mining offers a mechanism to mitigate conflicts among energy generators. By competing for hashrate and subsequent subsidy allocations, producers across all modalities can participate, provided they redirect surplus power to the grid during peak societal demand—a practice substantiated by operational precedents within Texas' ERCOT system and Georgia. Greater generation capacity enables producers to meet societal needs while retaining profitability through mining revenue. Notably, Bitcoin's agnosticism toward energy provenance ensures universal applicability.

This framework supports accelerated expansion of energy generation and distribution infrastructure by establishing a persistent,

---

[40] https://www.ercot.com/gridmktinfo/dashboards/

competitive demand profile—functioning as both initial and residual buyer. Such demand can leverage cost-effective resources or enhance existing operations for increased output and efficiency. The adaptability of this model accommodates diverse strategies, delivering a responsive load that stabilizes aggregate demand fluctuations across the grid—a transformative advancement in energy system dynamics.

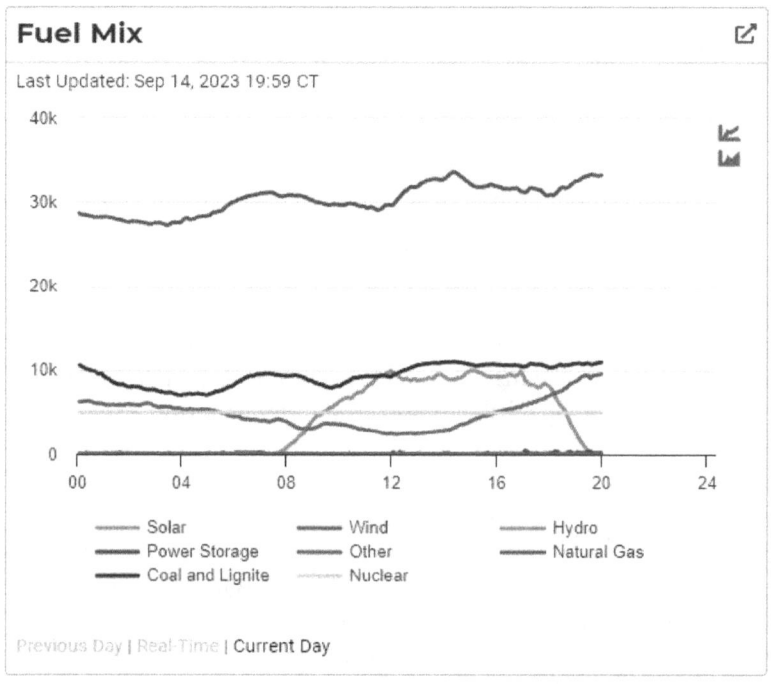

An optimally balanced energy system sustains a demand profile as stable as the consistent output typified by nuclear power generation. However, natural demand fluctuations (e.g., Figures 7 and 9) necessitate a flexible load to bridge disparities between supply and consumption. This requires a demand source capable of deactivation during unanticipated societal peaks, yet sufficiently advantageous—through operational enhancements and revenue generation—to warrant rapid reactivation once baseline requirements are met.

Bitcoin mining fulfills this role, as evidenced by operations within Texas' ERCOT system and Georgia. By absorbing excess capacity, miners stabilize demand variability while incentivizing generators to maximize production. This dynamic supports the development of infrastructure exceeding current requirements, establishing reserves for future utilization.

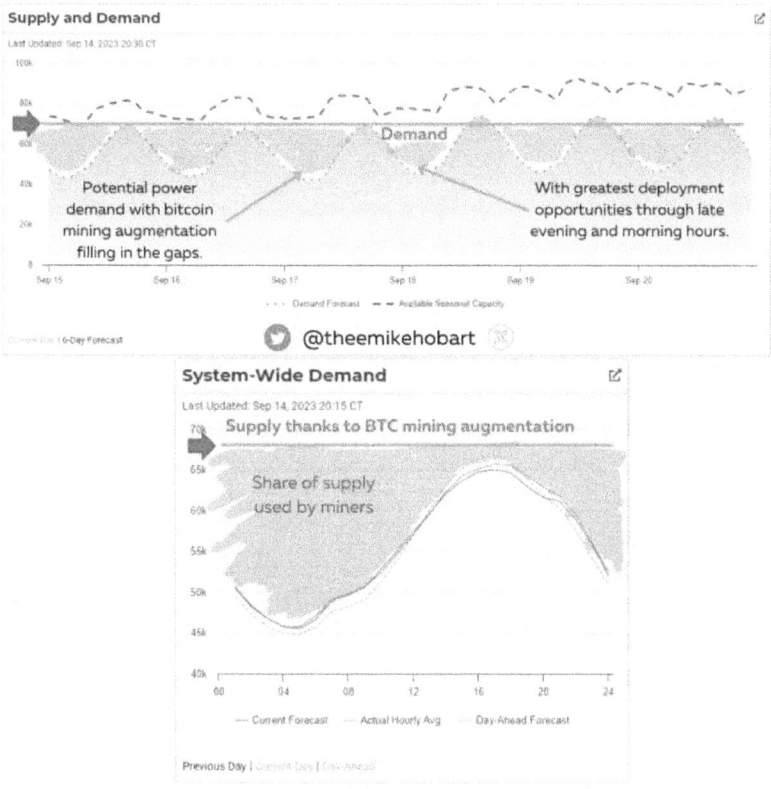

Figure 6 original version of graphics from: https://www.ercot.com/gridmktinfo/dashboards

When energy supply fails to streamline commodity production, demand persists unabated. Unlike gold or oil—where market dynamics moderate prices through increased production during high-price periods and reductions during lows—Bitcoin exhibits a distinct profile. Its consumption escalates with available energy inputs, constrained not by oversupply but by an intrinsic difficulty adjustment mechanism. As computational power intensifies and block completion accelerates, the network increases mining difficulty; conversely, it decreases when block production lags. This calibration precludes oversaturation, stabilizing supply irrespective of price fluctuations.

Mining pools further enhance this framework, enabling collaborative efforts to secure Bitcoin subsidies. Earnings are apportioned among participants proportional to their contributed computational effort, establishing a more stable revenue stream than solitary mining endeavors.

Conclusion

Energy generators across modalities stand to gain from integrating ASIC-laden datacenters to exploit the persistent demand inherent in Bitcoin mining. This competitive sector drives demand for enhanced chip efficiency and the utilization of abundant, underutilized energy capacity, prioritizing cost-effective sources. Consequently, producers and utilities leverage mining to optimize operational efficiency and augment revenue streams. These developments signal a reconfiguration of energy production paradigms, diminishing sectoral factionalism as generators align toward a unified, profit-driven objective.

Mike Hobart

CHAPTER 23: National Defense Through Bitcoin Mining

The United States confronts a complex maturation phase, contending with concurrent challenges of significant scale. Public health crises flourish, characterized by pervasive chronic metabolic diseases and escalating healthcare expenditures. Simultaneously, the banking sector undergoes pronounced consolidation amid contagion risks. Compounding these issues, an energy crisis—marked by concerns over generation capacity—persists, despite diminished public attention.

Recent events underscore this vulnerability. Apprehensions regarding sufficient energy production, briefly alleviated by a temperate European winter in 2022, resurface alongside documented grid infrastructure assaults along the Eastern Seaboard of the US prior to the 2022 holidays. These incidents highlight a critical question: how can grid power provision be fortified against such systemic weakening?

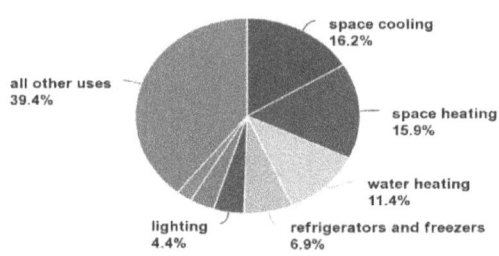

Figure 7 https://www.eia.gov/energyexplained/electricity/use-of-electricity.php

The solution currently available to us is two pronged:

First, the accelerated development of high-density energy projects—oil, natural gas, and nuclear power—is prioritized. Oil serves as a critical input for nuclear facility construction and the production of materials essential for upgrading and maintaining generation and distribution infrastructure, presently rated as deficient by the American Society of Civil Engineers[41]. Implementation faces extended timelines due to regulatory frameworks, safety standards, and bureaucratic processes, compounded by site planning, workforce allocation, and geopolitical realignments in manufacturing and labor, such as the relocation of semiconductor foundries to the United States. These factors necessitate meticulous coordination over protracted periods.

Second, Bitcoin mining leverages existing energy infrastructure. Facilities operating below capacity—a prevalent condition—are augmented with miners, providing economic incentives for consistent, predictable production across sustained durations[42]. This enables revenue generation during off-peak periods, optimizing

---

[41] https://infrastructurereportcard.org/cat-item/energy-infrastructure/
[42] https://papers.ssrn.com/sol3/papers.cfm?abstract_id=4634256

returns while mitigating equipment degradation through continuous operation[43]. Inactivity, a known contributor to corrosion and structural decline, is thus minimized, enhancing infrastructure longevity and resilience—an argument we made earlier with the Solow Model. Ancillary benefits include increased employment for maintenance, site management, and safety oversight. Additionally, miners' capacity for rapid deactivation—within 60 seconds, per Foreman Mining data—elevates Operational Readiness rates toward 90-100%, bolstering grid stability. This flexibility permits immediate demand curtailment, redirecting power to the grid during peak needs without compromising equipment integrity or incurring significant costs to producers or miners.

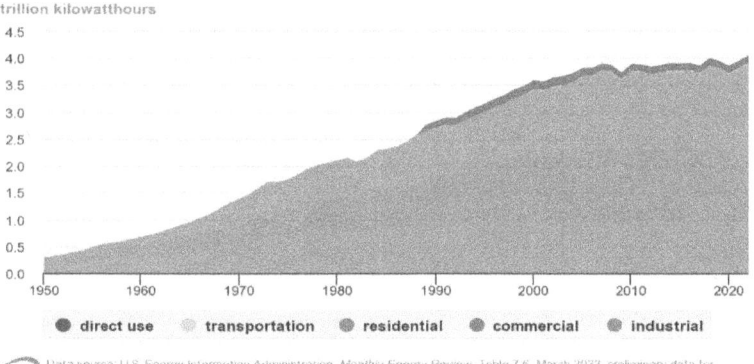

Figure 8 https://www.eia.gov/energyexplained/electricity/use-of-electricity.php

This approach advocates for the creation of what I call a "Rolling Reserve of Power," leveraging Bitcoin mining to elevate national or regional power production Operational Readiness (OR) rates to 90-95% utilization capacity—surpassing the current baseline of approximately 80%, as reported by Moody's. Contemporary systems exhibit fluctuations tied to peak demand periods, whereas this strategy stabilizes output consistency.

---

[43] https://papers.ssrn.com/sol3/papers.cfm?abstract_id=4899244

Current infrastructure maintains a deliberate capacity cushion, operating at 80% utilization (or 20% above designated full capacity) to accommodate extreme contingencies. Bitcoin mining integrates into this framework without compromising emergency demand response, as demonstrated in Texas' ERCOT system. Miners function as a regulatory mechanism, akin to a dam, modulating power allocation—retaining surplus during low demand and releasing it strategically during grid stress.

This capacity could mitigate risks of rolling blackouts amid near-term infrastructure vulnerabilities, while concurrently facilitating a synthetic expansion of generation capacity pending the realization of advanced solutions outlined in the first prong. Alternatively conceptualized, this reserve parallels a multi-barrel rotary weapon, sustaining continuous operational readiness for instantaneous power deployment as required.

Mike Hobart

CHAPTER 24: Decentralized Defense

What constitutes the foundation of a state currency? Potential underpinnings range from tangible assets (gold, oil) to coercive mechanisms (military force), demographic potential (labor or martial capacity), infrastructural outputs (energy production), or abstract assurances (future liberty). The inquiry extends beyond enumeration to the structural dynamics of economic systems.

Contemporary economies rely on interdependent sectors—finance, military, energy, agriculture, healthcare, and security—each providing specialized services for remuneration. Finance ensures transactional integrity through ledger management; the military secures territorial and strategic interests; energy sustains operational capacity; agriculture supports nutritional needs; healthcare mitigates symptomatic burdens; and security safeguards physical and digital assets. Bitcoin mining introduces a transformative overlay to this framework.

The Bitcoin protocol, through its incentive structure, recalibrates economic coordination. By establishing persistent demand—acting as both initial and residual buyer—it incentivizes maximal power production, yielding a stable revenue environment for energy providers. Integration with high-performance computing (HPC) for artificial intelligence initiatives diversifies income streams, enabling producers to command premiums. However, this revenue remains contingent on the operational stability of HPC-dependent entities, exposing producers to risks from governance or regulatory missteps—a game-theoretic consideration amplifying with scale.

Bitcoin mining mitigates these vulnerabilities, functioning as a stabilizing reinforcement for the energy sector. Its decentralized architecture precludes centralized errors, ensuring uninterrupted earning potential. Analogous to a military Quick Reaction Force (QRF)—units poised to support active operations—miners maintain readiness to curtail demand during grid stress, redirecting power to critical consumers. This flexibility contrasts with conventional curtailment participants, such as metallurgical operations (e.g., aluminum and steel smelters), whose prolonged shutdowns incur significant restart costs and production losses, propagating disruptions across dependent sectors.

This service enhances generation reliability and capacity, a foundation underpinning all economic activity. As adoption proliferates, economies increasingly rely on the Bitcoin mining-HPC nexus to justify expanded power production, embedding the sector's success within state interests. Coupled with Bitcoin's financial attributes—enabling wealth accumulation through savings resistant to debasement, these dynamic foster a decentralized defense of economic, monetary, and energy systems, aligning incentives across all participants.

CHAPTER 25: Ad Valorem

When tough times hit, give a quiet nod to Life for the nudge. Challenges are a chance to step up—proof we've got what it takes.

Lately, I've been picking my way through Frederic Bastiat's *The Law*. No rush, just savoring it. And wow, talk about timing. Those words feel like they were penned only yesterday. France in 1853 or here today, they still cut deep. If that doesn't make you wonder how far we've really come—or haven't—it's a wake-up call.

"Each of us has a natural right — from God — to defend his person, his liberty, and his property. These are the three basic requirements of life, and the preservation of any one of them is completely dependent upon the preservation of the other two."

Bitcoin's got this sacred spark. It guards your right to what's yours, backed by the toughest, mightiest computing web we've ever built. Cooked up by Satoshi Nakamoto himself. And it's all locked in by a handshake we've all nodded to, safe from meddling or tweaking. That's a rock-solid base for real, honest justice.

"If every person has the right to defend — even by force — his person, his liberty, and his property, then it follows that a group of men have the right to organize and support a common force to protect these rights constantly."

Here's the magic—this setup pulls folks together for something big: standing up for our rights. It's a shot at building a real land of freedom. Maybe that means shaking up old borders, or maybe it's tweaking what's already here once we wake up and see that our take on liberty's drifted off course. Either way, it's a fresh start.

> *"Force has been given to us to defend our own individual rights. Who will dare to say that force has been given to us to destroy the equal rights of our brothers? Since no individual acting separately can lawfully use force to destroy the rights of others, does it not logically follow that the same principle also applies to the common force that is nothing more than the organized combination of the individual forces?"*

It was this kind of thinking—these big, bold laws—that let a scrappy bunch of colonists take on a bigwig so out of touch, landing a punch for freedom that rippled 'round the world. And now, here we are, with the successors of those gutsy folks—who bet it all for liberty—turning around and snubbing that same freedom for others. Folks they figure are beneath them. These descendants grew up in cushy times, coasting along, forgetting what it took.

> *"If this is true, then nothing can be more evident than this: The law is the organization of the natural right of lawful defense. It is the substitution of a common force for individual forces. And this common force is to do only what the individual forces have a natural and lawful right to do: to protect persons, liberties, and properties; to maintain the right of each, and to cause justice to reign over us all."*

Mike Hobart

CHAPTER 26: Separating Money from the State

There comes a time, in the face of great discomfort and uncertainty, where one will be obligated to stand in defiance of a social contract with which one is in conflict.

Cultures and societies and governments can be rendered obsolete as our species and technologies advance. Progressing our understandings of where we find ourselves amongst one another. One might find the rules by which they abide no longer serve them, but constrain and oppress.

"Excess of liberty, whether it lies in state or individuals, seems only to pass into excess of slavery."

— Plato, *The Republic*

Today, the State enjoys this excess liberty.

Today, the State has unfettered access to our data.

Today, the State has the power of our time.

Today, the State has the power of our attention.

Today, the State has the power over our future.

The Second Renaissance

Over two-and-a-half centuries ago the world was saturated with authoritarian and aristocratic government rule. A select few chose to sacrifice the comforts of the now as a worthy cause. They saw the seemingly insurmountable task of attempting to secure their freedoms in the future if they did not take action. A future that was not a guaranteed victory. But a worthy cause, nonetheless.

Where a handful of American colonists stood tall against tyranny. As the rest of the world watched-on.

Where the few resisted the many. Against all odds and resources at their behest.

With sheer force of will, tenacity and spirit the underdogs refused to surrender as they knew they fought for what was right and what was true.

We find ourselves at a very similar crossroads. We do not know precisely where our world is heading. But we see many people across the world sacrificing their rights, and the rights of their children, in the search of immediate "security" and gratification.

We have an opportunity to stand, again, for what is right.

Only this time, we aren't limited to any one nation, or one geography. Through technological advancement we—as a people united on a global scale—can call for action taken together.

Only this time we can unite in a cause where all benefit, without need for violent revolution.

This time we cut the tether between money and state.

> *"Nothing else in the world… not all the armies… is so powerful as an idea whose time has come."*
>
> — Victor Hugo, *The Future of Man*

There comes a time... when an individual is presented with a choice; whether they play a role in history as progress is knocking at their door, or they resist change and defend "the old ways."

The decision to take the Orange Pill. Or the Blue Pill.

Will you stand for the right to shed the shackles of State power?

Will you stand and declare monetary independence from an unfair and broken system?

Will you choose to continue to let your hard-earned purchasing power be flippantly eroded away?

Or will you defend your time, your energy, your peace, and your future?

CHAPTER 27: The Looking Glass & The Sword

Carl Sagan had a knack for pinning down our world with a kind of clarity that can raise the hairs on the back of your neck. Looking ahead from his spot back then—our yesterday—he zeroed in on two big worries that still echo today.

"First…who is running the science & technology, in a democracy of people who don't know anything about it?

"The second… is that: science is more than a body of knowledge—it's a way of thinking. A way of skeptically interrogating the universe. With a fine understanding of human fallibility.

"If we are not able to ask skeptical questions; to interrogate those who tell us that something is 'true;' to be skeptical of those of authority, then we're up for grabs; for the next charlatan, political or religious, that comes ambling along."

Imagine a world where tech's pampered us—distracted us, nudged us into a comfy rut of ignorance. We're prostrated in a bed of comforts, of our own making, that would serve as our grave. Society needs a lifeline. A clear way out.

Thing is, folks like Sagan nailed the vibe but missed a beat. It's the money—the heartbeat of an economy—and the nudges it gives that shape how we act. Take a solid currency: it demands honesty, hard work. When purchasing power—be it ATP, gold, or Bitcoin—takes real grit to make, you can't just goof off. You've got to squeeze every drop of efficiency out of it, making the most of your time and hustle.

Flip that—when money's cheap, waste's no biggie. No slap on the wrist for flops or foolishness. If a kid acts up or a grown-up breaks the rules with no pushback, why change? It's a free-for-all—no stakes, no accountability. Picture Monopoly: if you can grab fistfuls from the bank while everyone else plays fair, no consequences, why quit? You wouldn't. Nature says you shouldn't. Out there, if you've got an edge, you ride it hard 'til the world catches up—or crumbles. Either the system finds balance, or it flops, clearing space for something smarter.

Smart folks sometimes miss this. It's playing out right now, just stretched over years, muddied by red tape and fancy finance tricks—what I'd call "economik" sleight-of-hand. Bitcoin cuts through that haze. It's a mirror to our soul, our economy—honest dealings on a tech backbone we can trust. Like Sagan said, science isn't just facts; it's a mindset. Bitcoin's our control group, a steady yardstick to measure the fiat mess against, testing what works, what flops, and how we grow. A real measuring tool for economies.

Here's the kicker: Bitcoin's not just a lens—it's a blade. It can slice through the waste, the rot, the junk piling up for over a century. But hold on—not a wild swing, like some Bitcoin fans might cheer for. I'm with Jeff Booth here: slow, steady, deliberate. Switching systems requires overlap—one fades, the other rises. This shift? It's the biggest we've ever seen, dwarfing the internet. Even paced out, it'll zip faster than history's ready for. If you've been in Bitcoin a while, you know—weeks feel like years, packed with more punch than the old world would witness in a decade.

Chapter 28: Bitcoin Enables Fiat Survival

While bitcoin may have been created with the aim to destroy fiat, I believe that in reality bitcoin will end up enabling fiat currency to survive, let alone flourish, into near-perpetuity.

Bitcoin will be co-opted by government(s) via allowing the digital asset to be custodied within traditional financial system. The FDIC has been working since spring 2021 to try and develop a framework to enable just this relationship. Meanwhile we have institutions like BNY Mellon rolling out custody solutions, and JP Morgan getting their trademark for a crypto wallet with fintech functionality.

By utilizing bitcoin, you have a digitally-native, scarce asset that is not limited by physical space. What this means is that there is a diminished need for physical space and effort dedicated to security or exchange. One of the major issues of using a gold peg is that as the coffers of gold grow, the need for physical space and services to maintain it, secure it, let alone exchange it, grows rapidly. By not being limited by space, or time (essentially), we effectively have an asset that can absorb the excesses of liquidity enabled by a pure fiat currency.

Mike Hobart

A Blackhole for Excess Liquidity

This asset operates like a black hole for currency; the more funds that get dumped into the asset does not change its state or nature. Making it perfect commodity for collateral, and the backing of a currency. It also solves the issue of trust due to the fact that there is no single individual or organization that exerts explicit control over the issuance, or operations, of the asset and it accompanying network. Could there be a future where bitcoin is transacted peer-to-peer (P2P) in de minimis amounts with the help of layer-x solutions, like the Lightning Network (LN)? Absolutely.

Markets have become aggressively financialized due to monetary debasement of fiat currencies, particularly the USD. Assets that, when financialized, produce referring complications. Assets that, in my opinion, have no business being financialized into investments, either entirely or at the very least to the extent to which they are financialized currently. For example one of my favorite points of this conversation; real estate.

Real estate and the ability to own a home has become so financialized and inflated due to excess liquidity sloshing throughout the market, trying to find a home. This has caused a demographic issue that is preventing generations of average citizens from being capable of owning a home. Then consider how this plies out effects into commerce: when there is less home buying, there is less structure to family formation, which not only directly means less citizens being born, but also much less wealth creation. Presumptively, whatever little "wealth" is created then gets gambled in over-inflated markets in the hopes of *making it big* in short time. Giving us not only wild volatility in markets, but also the over-inflation of collectibles like shoes and trading cards and digital images that can be screenshotted. This gambling activity causes wealth to accumulate to the most patient participants, which tends to favor those of already higher socioeconomic status. These individuals' ability to remain patient helps avoid falling prey to the most pervasive of emotions, fear. Fear of Loss (FOL) drives markets just as potently as FOMO (Fear of Missing Out). In all honesty, I would argue that FOL is more pervasive than FOMO. Ultimately resulting in an increasing widening of the wealth inequality gap.

With bitcoin as the base-layer asset, the excess liquidity can cease the violent sloshing activity reminiscent of an indoor pool during an earthquake. Making a home within bitcoin, over time, would then allow for not only more efficient means of value transfer but also credit deployment. With a pristine form of collateral, significant layers of risk can be eroded (including the obfuscation of risk through interpersonal secrecy, such as those deployed in the FTX-SBF debacle). Which is also a source of inefficiency within the current system; the reliance on bureaucracy and regulation. With something like bitcoin, automation can handle these dealings in seconds. Without needing to pay CPAs, lawyers, or administrators — which also comes with the costs of salaries, health benefits, commute to the office, etc.

Bitcoin is simply an iteration in innovation. Why fight something that improves efficiencies with reach of such depth and breadth?

Ultimately, it is my opinion that bitcoin will fail at destroying fiat. Banks & governments (both current and future) will have a use/need for credit and will utilize bitcoin for collateral. But it will succeed in enabling and enforcing responsibility and accountability. By utilizing both an asset that we know, without question, is finitely scarce as the foundation and alongside utilizing *responsible* credit we may get an economic system that *actually* makes sense.

> *"Actually there is a very good reason for Bitcoin-backed banks to exist, issuing their own digital cash currency, redeemable for bitcoins. Bitcoin itself cannot scale to have every single financial transaction in the world be broadcast to everyone and included in the block chain. There needs to be a secondary level of payment systems which is lighter weight and more efficient. Likewise, the time needed for Bitcoin transactions to finalize will be impractical for medium to large value purchases.*
>
> *"Bitcoin backed banks will solve these problems. They can work like banks did before nationalization of currency. Different banks can have different policies, some more aggressive, some more conservative. Some would be fractional reserve while others may be 100% Bitcoin backed. Interest rates may vary. Cash from some banks may trade at a discount to that from others.*
>
> *"George Selgin has worked out the theory of competitive free banking in detail,*

and he argues that such a system would be stable, inflation resistant and self-regulating.

"I believe this will be the ultimate fate of Bitcoin, to be the "high-powered money" that serves as a reserve currency for banks that issue their own digital cash. Most Bitcoin transactions will occur between banks, to settle net transfers. Bitcoin transactions by private individuals will be as rare as... well, as Bitcoin based purchases are today."

- Hal Finney, December 30th, 2010 — *Bitcointalk.org*

Chapter 29: Peaceful, Not Harmless

In 2023 a couple of friends approached me with an idea. Why not provide a space for the most patriotic of Americans—those who put their lives on the line for God and country—who are also on the future's edge? If the security of the freest of countries is a serious concern, then we are behooved to get as great a volume of soldiers exposure to bitcoin, in some responsible fashion, as fast as possible. It cannot be done without wisdom, and it must be done without greed or coercion. It must be done by choice, by free will. That is why we created the Bitcoin Veterans.

What started as a handful of crayon-eaters and grunts has grown into a network of intelligence and coordination that I could never have imagined. What these men have already done, and continue to accomplish, just because we planted a standard and provided a rally point, truly inspires.

OPN Bitcoin

Immediately, Operation Bitcoin (OPN Bitcoin) sprung up out of the organization. Providing an official funnel for our enlisted soldiers to be able to gain the understanding of an asset like bitcoin and its multiplicative mechanisms. OPN Bitcoin is further assisting our men-in-arms by giving them access to job-placement through the DoD's SkillBridge program. SkillBridge gives warriors that are looking at ETS'ing out of the armed forces exposure to work experience. By giving the soon-to-be veterans access to industry training, apprenticeships, and internships SkillBridge works to reduce the stressors in transitioning to civilian life and giving our warriors a mission right out of the gate. The value of a program like this cannot be understated. What civilians may have a hard time understanding is that the culture shock alone, going from military life—where everyone around you is of the same mind, operating under the same rules, where you know what to expect from the overwhelming majority—to civilian life is jarring. Veterans and enlisted experience frustration for multitudes of reasons.

Operation Bitcoin is a veteran owned and operated 501(c)(3) non-profit organization that is aimed at serving active-duty service members by providing Bitcoin education and creating pathways to employment within the Bitcoin industry. Through professional relationships, partnerships, and working with programs like SkillBridge OPN Bitcoin is providing veterans with a renewed sense of purpose and a new mission. Their close relationship to the Bitcoin Veterans also gives our vets an opportunity to find community with like-minded individuals that are also potently present within the bitcoin community and its industries.

The Brave Mission

One of my best friends, who just happens to be a co-founder of the Bitcoin Veterans, and the Bitcoin Veterans podcast, brought to me an idea immediately following the Bitcoin Veterans idea. An issue within our veteran population is the suicide rate. Every day we lose 22 men to suicide, and as Gabe Lord put it in our discussion at the

Unconference in Nashville in 2024, those are just the officially reported numbers. The reality is more than likely closer to 40 per day. Shane had the very clever idea of building a synergy between Bitcoin Veterans and assisting with the PTSD and mental health instability issue(s) that plague our fighting forces. The two of us had long been in discussion around the uses of psychedelics for therapeutic practices and just how foolish the institutionalized medical services and academia have become. We're already challenging centralized money and energy, why not add medicine to the list?

The Brave Mission is designed as a therapy program that is structured **around** the military man. Rather than try to force a vet into an environment that is foreign to him in order to get care, we place the vet in an environment that he is at home with. Instead of forcing a warrior into a walled environment like a doctor's office or a care facility, we take the man into the forest as a part of a squad.

One to two men, who're part of the Brave Mission, will be leading our struggling vet (or multiple vets depending on the comfort level of the client) through a psychedelic experience—we'll call them "journeymen." Meanwhile, there will be a secondary group who will be patrolling around the journeymen. These two groups encompass a complete experience for two different groups. While one is on their journey, the other is providing security. This allows for a return to the camaraderie that our warrior know and love, gives them separation from the civilian world that is a common stressor in the community, exposes them to the calming benefits of nature immersion, and gives them some great physical activity—as the area we have identified for these therapies is a forested mountainous region in the Appalachian Mountain Range. All of these aspects lower the guard of the struggling vet and bring him a level of comfort that will allow him to open up to men that he can confide in, far more than the environment of a cold and stuffy therapists office that wants him to "talk about his feelings." Furthermore, he has access to men who have had personal experiences with these difficulties and has been able to find success. This approach can be used as a PTSD treatment, a remedy for depression, or simply a retreat for the man that has found himself in a rut. We believe that this is a roadmap for success for our men that uphold the Red, White, and Blue.

Helene and North Carolina

The 2024 Nashville Bitcoin Conference was the starting gun. Hurricane Helene's destruction of North Carolina was the catalyst. The need of our fellow Americans was the fuel for something we could have never imagined.

A major part of the Bitcoin Veteran community is comms network that we have established. With multiple focuses we are capable of tackling a wide array of discussion topics. For example, when severe weather and natural disasters strike regions where we know we have Bitcoin Veterans (BVs), we are checking-in and gaining confirmation on conditions of our BVs and making sure there is no extraordinary need for the community. Hurricane Helene was one such extraordinary situation. We were gaining live updates from the ground thanks our BVs, especially the situation reports (sitreps) on the lack of availability for resources and the breakdown of logistics channels. Multiple BVs shared that they were going out into the field themselves to provide aid. These men were our ADVON (Advanced Echelon), gaining intimate knowledge of the situation on the ground and reporting needs by priority.

While gaining regular updates from our ADVON myself and a team of 1-2 dozen others volunteered nearly all our free time towards aggregating intel on the situation (supply needs, supply drop location needs, medical emergencies, available units for dispatch, etc.), compiling them into shareable documents, and dispersing the information to as many like-minded groups as we could come into contact with. We had the majority of our people operating during the day and I was working to connect networks and compile more intel throughout the night (since I worked nightshift at the time). We quite literally had 24-hour coverage of the situation.

Our ADVON element made contact with a group called the Christian Rangers, a group aimed at providing aid for such situations here at home and abroad, while being the ones to help set up helicopter landing zones (HLZ) like HLZ ZOMBIE—which ended up serving as a major logistical node for rescue and recovery efforts for the duration of the crisis. We decided that our efforts would go to the best ends by working synergistically with the Christian Rangers to combine force multiplication efforts. After the first ADVON push several of our men stated that they would be going back out to help, at which point my friend and co-founder Jordan Gambrell and myself, decided we would also deploy.

We received numerous reports of dangerous conditions. From the lack of utilities and resources causing desperation, to cartels like Tren de Aragua (TdA) taking advantage of the situation to kidnap women and children amidst the chaos of the disaster, chemical spills from factories in the mountains and illegal drug operations, to militia groups being forced out of refuges hidden in the Appalachia Mountains, there was no shortage of chaos. While Jordan devised a packing list, we took to social media to offer up our time to gather donations to deliver to the people of North Carolina. What followed was beyond miraculous.

In just a few weeks the Bitcoin community had provided us with $120k in donations. We were receiving updates while we were out in the field using the first flow of funds to purchase generators, fuel, chainsaws, cold and wet weather gear, and comms equipment. At the same time, we were sharing updates with what we were accomplishing and relaying the reality of the situation that was not making it to the outside world. We ended up being contacted to be a part of an X (Twitter) Spaces with the Mario Nawfal team to share updates and make calls to duty for those that were willing and able. The Christian Rangers were the biggest part of our efforts, and we used our social media presence to push them into the well-deserved spotlight. I aided the Commander of the Christian Rangers and a few of their rangers to navigate the workings (and bugs) of the Spaces platform. To be able to help our brothers and sisters at the Christian Rangers get in front of over 1 million eyes, and over 550k listeners will forever be a personal achievement of mine.

Aside from providing funds, resources, and manpower, our other focus was in preparing the local leaders and community to be self-sustaining. We aimed to provide them with a hand to get back on their feet, but we wanted them to primarily **get on their feet**. We led several key leader engagements to discuss preparation needs and direct the community in an organized fashion towards mutual ends.

We ended up with more donations than we knew what to do with, the greatest need by the time we got into the area was brawn and brains. Being a group of crayon-eaters and Army grunts, we definitely had no shortage of brawn. The brains however... that's up for interpretation. The remainder of our funds went to Resilient Recovery, who went on to gain utilization of a massive warehouse space and provided continued efforts through the winter and into the spring. All thanks to the support of the bitcoin community.

Our BVs

What surprises me most of all about what we have going on at Bitcoin Veterans isn't what I have discussed thus far. What inspires me so much about our BVs isn't the disaster relief efforts, or the interest in providing PTSD treatments to our brothers, but what these guys do on their own once we provide them an environment to muster.

The Second Renaissance

We have BVs putting on their own conferences like Bitcoin Alaska. We have men who are working in the mining sector with projects like Supporting Pleb Miners, HeatPunks, and several Open-Source projects. We have gentlemen working on building a CIVNET, where civilians who have an interest in participating in the Bitcoin Veterans mission can rally, as well as a mining podcast coming. We have BVs leading their own cryptography podcasts, speaking at developer conferences, as well as leading the DC BitDevs.

The Bitcoin Veterans have already come so much farther than I could have irresponsibly dreamed up, that there is no knowing where we are going next. I hope to one day have a physical security service like our friends over at Paladin Tower Tactics, but I might leave that to someone else to lead.

Mike Hobart

The Orange Sun Rises

Ultimately, the goal of this publishing was multipronged.

Identify the greatest sources of our issues. As most of our population prefers to hone-in on the symptoms rather than the sources of our dysfunctions. The centralization of information access and sourcing, over-reliance on appeals to authority, outsized centralized economic control—which allows for manipulation of information gathering and spread, food sourcing and taking care of our land, the ways that economic factors which lie downstream of previously mentioned relationships affect very important aspects of individual health, and power generation.

The best part about our situation is providing a savings tool for the average citizen, while also tackling power provision gets the ball rolling on seeking the solutions for all the others. Fix the power problem and enable The People to establish financial security. Strengthening the farmers, the welders, the plumbers, the homebuilders, the soldiers, the warriors, the fathers, the mothers, the sons and the daughters. The core of the country, and the future of us all.

The Second Renaissance

While this is all occurring, we will be seeing **rapid** dissemination of information that has been gatekept from the public to charge a toll—for profit. We are already seeing this happening. The ivory towers are taking fire, while they attempt to foolishly defend the worlds of yesteryear, where they held power. We are in for a rapidly changing world, much of it for the better. But we will see the volatility continue to swell for several years. Freedom will continue to need defending and to be fought-for. Freedom money and freedom tech like Meshtastic will enable it. And for that food and energy will be required. See how this all reverberates back and forth? That is why these messages resonate.

I may have started out scaring you, but I needed to. You need to be scared seeing how absolutely busted our world is. The cracks have been spotted. Now the sense of urgency smoldering within your soul can push you to get to work. Just like it did within me. We all need it from time to time.

Now…go forth and conquer.

Mike Hobart

Thank You

I am immeasurably thankful to you for reading my work here. I hope I have had at least one impact on the ways in which you are viewing the world. I hope that I have helped highlight at least one reason for you to be optimistic for your future. What all too often gets lost is the reality of the Yin and Yang. We must have darkness and hard times in order to build and then appreciate the good times.

We have many, many tools and information at our disposal to find solutions and receive solutions from far more nubile minds than ever before in history. All we need is to get the tools, the internet, and the funds to these individuals to get started. Then we let the market do its thing, and we will get the solutions to every problem we have.

We are on the eve of revolutions in everything meaningful. Don't let that kind of magic get ruined by losing long-term gains in the chase of short-term gratification.

I want to hear what you thought of the book, you can find me on Twitter (@theemikehobart) and on NOSTR.

REFERENCES:

Booth, Jeff. *The Price of Tomorrow: Why Deflation Is the Key to An Abundant Future.* Stanley Press, 2020.

Brill, Steven. *Tailspin: The People and Forces behind America's Fifty-Year Fall--and Those Fighting to Reverse It.* Alfred A. Knopf, 2018.

Epstein, Alex. *Fossil Future: Why Global Human Flourishing Requires More Oil, Coal, and Natural Gas--Not Less.* Portfolio/Penguin, 2022.

Kelly, Kevin. *What Technology Wants.* Penguin Books, 2014.

Montgomery, David R. *Growing A Revolution Bringing Our Soil Back To Life.* W.W. Norton & Company, 2017.

Citations: *presented in order of reference within text.*

1. "National Diabetes Statistical Report." *CDC.Gov*, 15 May 2024, www.cdc.gov/diabetes/php/data-research/index.html.
2. "ASCE's 2021 American Infrastructure Report Card: GPA: C-." *ASCE's 2021 Infrastructure Report Card |*, 24 May 2024, infrastructurereportcard.org/.
3. Reich, Charles. "The New Property." *The Yale Law Journal*, 25 Nov. 2021, openyls.law.yale.edu/handle/20.500.13051/14924?show=full

4. Nicholson, Christie. "How Does Due Process Protect a Public Employee? - Findlaw." Findlaw.Com, 15 Jan. 2024, www.findlaw.com/employment/wages-and-benefits/how-does-due-process-protect-a-public-employee.html
5. "Graveyard of Empires - Dirt: The Erosion of Civilizations - David R. Montgomery." *Publicism*, 2016, publicism.info/environment/dirt/4.html.
6. Reader, Uncle John's Bathroom. "Eureka! The Discovery of Photosynthesis." *Today I Found Out*, 20 July 2015, www.todayifoundout.com/index.php/2015/07/eureka-the-discovery-of-photosynthesis/.
7. Psychology, Department of. "Consistent Sex Differences in Cortisol Responses to...□: Biopsychosocial Science and Medicine." *LWW*, Nov. 1992, journals.lww.com/bsam/abstract/1992/11000/consistent_sex_differences_in_cortisol_responses.4.aspx.
8. Sherman, Gary D., and Pranjal H. Mehta. "Running Head: Stress, Cortisol, and Social Hierarchy." *UCL*, discovery.ucl.ac.uk/id/eprint/10085954/1/ShermanMehta_StressCortisolHierarchy.pdf. Accessed 2024.
9. Baeken, C., et al. "Cortisol Response to Stress: The Role of Expectancy and Anticipatory Stress Regulation." *Cortisol Response To Stress: The Role of Expectancy and Anticipatory Stress Regulation*, Science Direct, 25 Oct. 2019, www.sciencedirect.com/science/article/abs/pii/S0018506X18304860.
10. Dekkers, Tycho J., et al. "A Meta-Analytical Evaluation of the Dual-Hormone Hypothesis: Does Cortisol Moderate the Relationship between Testosterone and Status, Dominance, Risk Taking, Aggression, and Psychopathy?" *Neuroscience & Biobehavioral Reviews*, Pergamon, 7 Dec. 2018, www.sciencedirect.com/science/article/pii/S0149763417306784.
11. Crewther, Blair T, et al. "Serum cortisol as a moderator of the relationship between serum testosterone and olympic weightlifting performance in real and simulated

competitions." *Biology of Sport*, vol. 35, no. 3, 2018, pp. 215–221, https://doi.org/10.5114/biolsport.2018.74632.
12. Herre, Bastian, et al. "Marriages and Divorces." *Our World in Data*, 25 July 2020, ourworldindata.org/marriages-and-divorces#divorce-rates-increased-after-1970-in-recent-decades-the-trends-very-much-differ-between-countries.
13. Linville, Tani M. *Project MKULTRA and the Search for Mind Control: Clandestine Use of LSD Within the CIA*, Cedarville University, 26 Apr. 2016, digitalcommons.cedarville.edu/cgi/viewcontent.cgi?article=1005&context=history_capstones.
14. Katara, Harsh. "Money vs Currency." *Top 6 Differences (with Infographics)*, WallStreetMojo, 19 Jan. 2019, www.wallstreetmojo.com/money-vs-currency/.
15. American Society for Biochemistry and Molecular Biology. "How ATP, molecule bearing 'the fuel of life,' is broken down in cells." ScienceDaily. ScienceDaily, 9 March 2010. <www.sciencedaily.com/releases/2010/03/100301091428.htm>.
16. The Editor's of Encyclopedia Britannica. "Adenosine Triphosphate." *Encyclopædia Britannica*, Encyclopædia Britannica, inc., 24 Feb. 2025, www.britannica.com/science/adenosine-triphosphate.
17. Fisheries, NOAA. "Impacts of Invasive Lionfish." *NOAA*, www.fisheries.noaa.gov/southeast/ecosystems/impacts-invasive-lionfish. Accessed 24 Mar. 2024.
18. Ruiz-Ojeda, Francisco Javier, et al. "Effects of Sweeteners on the Gut Microbiota: A Review of Experimental Studies and Clinical Trials." *Advances in Nutrition (Bethesda, Md.)*, U.S. National Library of Medicine, 1 Jan. 2019, pmc.ncbi.nlm.nih.gov/articles/PMC6363527/.
19. Wells, Katie. "How Artificial Sweeteners Affect the Body." *Wellness Mama®*, 17 Feb. 2020, wellnessmama.com/health/artificial-sweeteners/.

20. Huberman, Andrew. "Dr. Andy Galpin: How to Build Strength, Muscle Size & Endurance." *YouTube*, Andrew Huberman, 28 Mar. 2022, youtu.be/IAnhFUUCq6c?si=Vr37WJ2JVevWa21H.
21. Rodriguez, Tori. "Exploring the Link between Persistent Infection, Inflammation, and Mood Disorders." *Psychiatry Advisor*, Psychiatry Advisor, 25 May 2018, www.psychiatryadvisor.com/features/exploring-the-link-between-persistent-infection-inflammation-and-mood-disorders/.
22. Swithers, Susan E. "Artificial Sweeteners Produce the Counterintuitive Effect of Inducing Metabolic Derangements." *Trends in Endocrinology and Metabolism: TEM*, U.S. National Library of Medicine, 1 Sept. 2013, pmc.ncbi.nlm.nih.gov/articles/PMC3772345/.
23. "Jiaozi (Currency)." *Wikipedia*, Wikimedia Foundation, 29 Jan. 2025, en.wikipedia.org/wiki/Jiaozi_(currency).
24. Peden, Joseph R. "Inflation and the Fall of the Roman Empire." *Mises Institute*, Mises Institute, 21 Dec. 2020, mises.org/mises-daily/inflation-and-fall-roman-empire.
25. Gladstein, Alex. "Uncovering the Hidden Costs of the Petrodollar." *Bitcoin Magazine*, Bitcoin Magazine, 28 Apr. 2021, bitcoinmagazine.com/culture/the-hidden-costs-of-the-petrodollar.
26. Fuhrman, Joel, et al. "Changing Perceptions of Hunger on a High Nutrient Density Diet - Nutrition Journal." *BioMed Central*, BioMed Central, 7 Nov. 2010, nutritionj.biomedcentral.com/articles/10.1186/1475-2891-9-51.
27. Atlas, Ronald M, and Terry C Hazen. "Oil Biodegradation and Bioremediation: A Tale of the Two Worst Spills in U.S. History." *Environmental Science & Technology*, U.S. National Library of Medicine, 15 Aug. 2011, pmc.ncbi.nlm.nih.gov/articles/PMC3155281/.
28. Vuorinen, Heikki S, et al. "(PDF) History of Water and Health from Ancient Civilizations to Modern Times." *History of Water and Health from Ancient Civilizations to Modern Times*, Research Gate, Mar. 2007, www.researchgate.net/publication/250142768_History

_of_water_and_health_from_ancient_civilizations_to_modern_times.

29. Ferguson, Roger W, and Upamanyu Lahiri. "The History and Future of the Federal Reserve's 2 Percent Target Rate of Inflation." *Council on Foreign Relations*, Council on Foreign Relations, 15 June 2023, www.cfr.org/blog/history-and-future-federal-reserves-2-percent-target-rate-inflation-0.
30. Hobart, Mike. "Bitcoin Mining: Power Games." *Bitcoin Mining: Power Games*, Simply Bitcoin Unfiltered, 5 Apr. 2023, simplybitcoin.substack.com/p/bitcoin-mining-power-games.
31. Hill, Kyle. *We Solved Nuclear Waste Decades Ago*, YouTube, 27 Mar. 2022, youtu.be/4aUODXeAM-k.
32. "Why Your Electric Bill Is So High — and Could Keep Climbing | WSJ." *YouTube The Wall Street Journal*, The Wall Street Journal, 10 May 2022, youtu.be/fFXcBIFoxdE.
33. DeSilver, Drew. "For Most U.S. Workers, Real Wages Have Barely Budged in Decades." *Pew Research Center*, Pew Research Center, 7 Aug. 2018, www.pewresearch.org/short-reads/2018/08/07/for-most-us-workers-real-wages-have-barely-budged-for-decades/.
34. Elliott, Tom. "Rep. @RashidaTlaib Challenges Bank CEOs to Agree to Stop Funding Fossil Fuels, Is Rejected by Every Single One | X.Com." *X.Com*, X.com, 21 Sept. 2022, x.com/tomselliott/status/1866908469535572350.
35. Gallagher, Tom. "PALEOCLIMATOLOGY Part 1." *YouTube*, Roger Palmer, 19 Nov. 2020, youtu.be/K6tWEjkEiZU?si=AaRvdsq8v0XKD2wP.
36. Gallagher, Thomas. "PALEOCLIMATOLOGY Part 2." *YouTube*, Roger Palmer, 3 June 2021, www.youtube.com/watch?v=iZSYSWPYEbU.
37. Gallagher, Thomas. "Paleoclimatology Part 3." *YouTube*, Roger Palmer, 28 July 2021, www.youtube.com/watch?v=YMHKt9ylPpQ.
38. Stern, David I, et al. "The Impact of Electricity on Economic Development: A Macroeconomic

Perspective." *eScholarship, University of California*, 31 May 2019, escholarship.org/uc/item/7jb0015q.
39. "The Solow Model and the Steady State." *YouTube*, Marginal Revolution University, 12 Apr. 2016, youtu.be/LQR7rO-I96A.
40. Stern, David I, and Astrid Kander. "The Role of Energy in the Industrial Revolution and Modern Economic Growth." *Archive.Org*, 2012, web.archive.org/web/20240301023434/https://crawford.anu.edu.au/distribution/newsletter/research-newsletter/pdf/Energy-Journal-Stern.pdf.
41. "Grid and Market Conditions." *Electric Reliability Council of Texas*, www.ercot.com/gridmktinfo/dashboards/.
42. "U.S. Energy Information Administration - EIA - Independent Statistics and Analysis." *Use of Electricity - U.S. Energy Information Administration (EIA)*, 2023, www.eia.gov/energyexplained/electricity/use-of-electricity.php.
43. "Energy." *ASCE's 2021 Infrastructure Report Card |*, 12 July 2022, infrastructurereportcard.org/cat-item/energy-infrastructure/.
44. Carter, Nic, et al. "Leveraging Bitcoin Miners as Flexible Load Resources for Power System Stability and Efficiency." *SSRN*, 30 Nov. 2023, papers.ssrn.com/sol3/papers.cfm?abstract_id=4634256.
45. Rudd, Murray A., and Dennis Porter. "Economic Integration of Bitcoin Mining in Renewable Energy and Grid Management." *SSRN*, 24 July 2024, papers.ssrn.com/sol3/papers.cfm?abstract_id=4899244.
46. "U.S. Energy Information Administration - EIA - Independent Statistics and Analysis." *Use of Electricity - U.S. Energy Information Administration (EIA)*, 2023, www.eia.gov/energyexplained/electricity/use-of-electricity.php.

www.ingramcontent.com/pod-product-compliance
Lightning Source LLC
Chambersburg PA
CBHW070457100426
42743CB00010B/1663